Alchemy Unveiled

Alchemy Unveiled

For the first time,
the Secret of the Philosopher's Stone
is being openly explained

by

Johannes Helmond

.

.   .

fr. R + C

By decree of the Order
of the
hermetic initiated
Gold- and Rosicrucians
of 1710

Monte Abiegno,
MCMLVII, Sol 18° gemini.

Translated into English and Edited by
Gerhard Hanswille and Deborah Brumlich

Merkur Publishing Co. Ltd.
Scarborough – Canada

Canadian Cataloguing in Publication Data
Helmond, Johannes
Alchemy Unveiled

Translation of Die entschleierte Alchemie.

Exclusive permission to translate Johannes Helmond's Book:
"Die entschleierte Alchemy" (1963) was granted to Merkur Publishing Co. Ltd.,
Scarborough – Ontario, Canada by the Publisher Rohm Verlag, Bietigheim.

ISBN 0 9693820 4 9

Printed in Canada by Coach House Press, Toronto, Canada.

Contents

*The publishers would like to thank Franca Gallo
for the many long hours of dedicated work
in the preparation of this book.*

Introduction

Just a few decades ago, Alchemy was belittled as medieval superstition. However, since then, scientists have been successful by means of atomic-fission, to produce a transformation of the elements. Now, at last, the possibility of Alchemy in Principle has to be admitted. In our present times, psychology began to embrace Alchemy and reached a certain level of understanding, whereby they came to the conclusion that Alchemy was less a process of physical transmutation, but more a process of an inner psychic transformation of a Human Being. On account of this conclusion, Alchemy is now being looked upon in a different light.

It should, however, be pointed out that modern psychology, especially as it is represented by C.G. Jung, did not grasp the central problem of Alchemy. C.G. Jung conceives the process of Alchemy to be solely and totally as "psychological" and looks upon it as just a kind of "Individuation-process"; that means, an integration of the content of the unconscious. However, Alchemy is considerably more than that. Jung and his fellow psychologists overlook hereby the eminent realistic character of Alchemy, which is interdependent with the "inner body," the "corpus subtile" of the Human Being. Jung, however, considers this to be the purely, psychological collective unconscious. Even though, upon closer examination, the substantiality of this "Corpus sideribus," as it was named by Paracelsus, cannot be doubted.

Concerning the question of Alchemy, we are not dealing with a mere psychoanalytical process, nor are we dealing with a simple "Cooking of Gold" which it is deemed to be, by many. The "Lead" which the Alchemists are changing into Gold is not the Pb of the common chemists, but instead, is the dark Saturnus of the hermetic Philosophers. Again, this is not the external "celestial or heavenly body" but instead, is its inner astral principle, which is present everywhere, in the materia.

9

When we are dealing with Alchemy, we are not dealing with the processes of the great Outer-World, but instead, we are dealing with the very secretive microcosmic processes, which form, since time immemorial, the central subject of the mysteries. Therefore, it can be said, and rightfully so, that the hermetic sciences are the oldest sciences mankind has. It is probable that this science originated from a tradition, to which even the ancient Egyptians were in awe of and they depicted them as "Secrets of primeval Times."

The Greeks and Israelites received these mysteries from the Egyptians and they were well-kept by them, especially by the School of the Prophets, which became extinct during the "babylonian Captivity." Later on, this ancient hermetic tradition was taken up again, by original Christianity and it was continued by the first Disciples and Apostles as "Disciplina arcani." The hermetic science came to the New Platonists and to the arabian Philosophers from the original christian Nazarenes and through them, it reached Europe in the middle ages.

As a result of this, several hermetic Centres were founded in Europe. That is how in 1429 the "Order of the Golden Fleece" came into being. Out of this emerged the Rosicrucians, who did not initially appear publicly, but remained very private.

It was not until after the 30-year war, that the Rosicrucians established a closer link with the rest of the world, especially in the form of the purely hermetic Gold- and Rosicrucianism of 1710. This tradition is being continued in the present consciously, only by our Order.

In our present times, there are writings carrying the Name of Alchemy, which are provided by certain sources and presented in a very appealing cover. These writings caused, in many instances, a considerable misuse, as they contained the most absurd theories. Our illuminated Order is seeking to redirect these false doctrines through the publication of this treatise. It is not the intention of our Order to disclose all the Secrets of Alchemy and to throw them under the feet of the great crowds of the curiosity seekers, speculators and sceptics. Instead, it is meant to show the few sincere Seekers of the hidden Wisdom, the right direction, where one must search for the great Elixir. These Seekers will then have a measure, a standard in their possession, to assist them to differentiate between the writings of the true hermetic Adepts and the elaborations of the dreamers and charlatans.

Whosoever, of the sincere Seekers and Students of the hermetic sciences is called upon, will without a doubt, sooner or later, be in touch with our illuminated Order and will receive further information and guidance, which are not contained in these writings before you.

Prologue.

The Main Purpose and Finis Ultimus of the Hermetic Art is not to produce Gold, as it is the belief of the ill-informed lovers of Gold, but instead the study of God's beautiful Miracles, which lie hidden in the abstruse rerum centro, and to contemplate the Sacrarium Naturae benedictae, ejusque Majestatem occultum remoto velo.

– (Axiomata, 1736)

It is well known that the Alchemists, since time immemorial, made use of a secret code, the key to which was only in the possession of the Initiates. That is why Herman Fictuld said, only the true Adepts understand the writings of the genuine Hermetics.

In fact, it is totally impossible for outsiders to decipher the hidden meaning of the Old Writings of the Adepts. Many have been mislead due to their greed for gold or the wonderful effective powers of the life elixirs. They experimented blindly, but their harvest was no more than great losses and disappointments. Whosoever does not understand the right Theory, will hardly advance to the true Practice of the Philosopher's Stone, and whosoever does not know what he is searching for, also does not know what he will find.

True Alchemy, in reality, is a quabbalistic Art which requires a patient examination of the genuine hermetic writings and a deep-founded study of Nature. It also requires a revelation, either through an initiated Adept, or through an inner divine illumination.

The style of the true writings of the Adepts has always been obscure and parabolic, especially where the Sages express the Hermetic Art in allegorical figures and puzzles. That is where they express themselves in the clearest manner, whereas when

13

everything seems to be clear and understandably written, the Art is hidden the most.

The reason for this is to keep unsuitable elements away from the Hermetic Art from the beginning, thereby preventing any misuse of same. If the Initiates were to unveil all of the Mysteries of Alchemy indiscriminately, then the ignorant would treat the Art with contempt, because they did not understand it, and for those who do have the understanding, it would be considered too easy. However, when the deeper truth is veiled in an allegorical envelope, then the deeper truth is safe from disrespect and it guarantees the diligent ones the incentive to philosophize.

Under certain circumstances, this mystical veil can be taken too far, so that even the sincere seeker will not be able to explore the allegories. It is therefore a serious commandment of the hour, to acquaint those who truly strive with the true matter of the Philosopher's Stone, and with the key of the language of the hermetic art, to avoid being totally mislead.

The one who first understands a true Hermetic, will learn, little by little, to understand all true Adepts.

In order to give the serious Seekers of Wisdom a well-meaning piece of advice, these writings should lift the veil which the Adepts of old spread over the Hermetic Art, only to the point permitted by the unwritten laws of the Mysteries, so that the sincere Seekers of Truth will not miss the right path to the Lapis Philosophorum.

At the same time, through this veil, the deceitful "Goldmakers" and the lost "tincture-cooks," who up to now only misused the name of Alchemy, will finally be entirely lead ad absurdum (towards nonsense). In the "microcosmic Prologue" of 1720, this is what is said about the Pseudo-Chemists:

> The unworthy Lovers and Seekers, with their perverse or
> unnatural Subiectis and Laboribus, cannot bring the philo-
> sophical Sun and Moonchild into the World. They seek

this divine natural Art without the recognition of God and Nature; they do not even know what kind of thing Nature is, and they know not the inner workings of Nature. Their minds go constantly around the circumference, and they speak of the many effects of Nature, but they do not find the Centre of Nature, which causes all of the effects of Nature.

They should draw out of the living fire, out of the living metals of the Sages, the seed; consequently make the Mercury through the Mercury, or make the first materia through the first materia. They do not know what the Life and the seed of metals is, nor do they know what Mercury and the First materia are. Instead, they work in dead and deceased subiectis, such as common Gold, Silver, and Mercury, and they also do this with Wood, Coal, Lamps, or with some kind of deadly fire, being of the opinion that through this, they will prepare a life-giving and life-extending Universal Remedy and Tincture; as if Life and Death were in their hands. They operate with nothing but bodies, whereas Nature deals with nothing but spermatic things. They seek an easy Art and heavy labour, but instead, it is a difficult Art and easy labour. They spend great amounts of money for materials which can be had for nothing (they cannot be bought in a store or Apothecary); they must be taken directly out of Nature.

– *(Microcosmic Prologue, 1720)*

The Symbolic Language of Alchemy.

"It is impossible, that any Mortal understands this Art, unless he has been previously enlightened by the Divine Light." — *(Dorneus, Theatr. Chem. 1602)*

The terminology of the old spagyric Adepts is indeed a dark labyrinth for all those, who do not thoroughly know the principles of the hermetic sciences.

In order to obtain this knowledge, many years or even decades of study of the old classic Writings of the Adepts is a prerequisite. In addition, there must be a higher enlightenment or instructions given by a true Master of the Art, before you can truly master the Theory of the hermetic sciences. Without this, no one should dare to approach the Practice of the Hermetic Art.

This is why, this is said in the microcosmic Prologue:

"It is with the Exegesis of the Writings of the Sages, as it is with the Bible: whosoever wants to understand and interpret both must ask **GOD** for the **SPIRIT** and the **LIGHT**, out of which and in which they are written; he must simply have the **LIGHT**, which shone for the Sages, otherwise he would judge this, like a blind man would judge colours.

Alas those, who have not written out of the Fountain of Nature, but instead out of processes or out of compiled philosophical books, are seducers, because they recite truth mixed with error, to blind the reader and the innocent seeker of the pure Light. Those, who understand the matter thoroughly, recognize clearly in all books, the truth from the lies; the pure Spirit of Nature dictates more to him, than what he can write."

— *(Microcosmic Prologue, 1720)*.

The uninitiated encounters the most difficulty in the multitude of meanings of the individual alchemistical symbols and their synonyms.

Mercury has often given rise to confusion, because you have to know precisely the difference between the common philosophical Mercury (mercurius universalis) and the "Mercury of the Philosophers" (Mercurius Philosophorum), just as the simple mercurius corporeus should not be mistaken with Double Mercury (Mercurius Duplicatus). In addition, the Alchemists also knew the difference between a white and a red, a fleeting and a fixed mercury; apart from the fact that Mercury would occasionally be considered to be a Sulphur or Salt.

Only the truly informed have the ability to recognize certain marks or signs to indicate which mercurius is specifically spoken of in the old Writings of the Adepts. It is, therefore, not surprising when famous researchers, such as the well-known Swiss Psychologist, C.G. Jung, could not find his way among the multitude of alchemistical Mercury-Symbols, because he did not possess the key to the old hermetic symbolic language. Without the key, Alchemy remains an obscure puzzle.

The Hermetics knew of, besides the ambiguous Mercury, various kinds of Sulphur and Salt; at least three different kinds of Gold; four kinds of Fire; three kinds of Water; several kinds of Earth; and three different Materia for the Great Work. The Initiate must know how to distinguish all of them from each other.

This multiplicity could deter many friends of the hermetic Mysteries, however, the ones who sincerely search will not be discouraged through the hardships of searching, since the reward of such a search is truly not a small one!

It is therefore prudent (necessary) to truly master the Theory of the Hermetic Sciences before beginning the alchemistical Practice. You must have a total understanding of the old Writings of the Adepts and a deep insight into the matter itself.

17

Above all, it is absolutely necessary to have a thorough knowledge of the Hermetic Theory of the Elements, which must stand at the beginning of all alchemistical contemplations. This is what Eugenius Philaletha had to say:

> "Whosoever wants to obtain a benefit in the Art must learn to know the elements and their effects before they seek the Tincture of the Metals, observe their effects and reproduce them." — *(Eug. Philaletha in "Euphrates")*

The four Elements of Alchemy are: Fire, Air, Water and Earth. They are in no way identical to the well-known appearances of same. The Earth of the Alchemists is not the coarse impure Earth that we step upon with our feet, but instead the pure Element of the same, the central fixed Nitrum, which contains in itself the Nature-Sulphur, through which she coagulates the Mercury.

When the Hermetics speak about their Water, they are not speaking of the H_2O of the Chemists, but instead of the "Water of the Sages" in which they cook their Fire (their Sulphur).

The common Chemists cook their water and their materials in the fire; the hermetic Sages cook their Fire in their Water!

> "Everybody knows how to cook water in the fire, however, if they would know how to cook our FIRE in our WATER, then their Knowledge of Nature would rise above that of the cooks." — *(Thomas Vaughan, Aula Lucis)*

"Combure in acqua, lava in igne!" is therefore a well-known Phrase of the Alchemists. It seems to sound paradoxical, but, Wisdom is always a paradox.

The Air of the Alchemists is also substantially different than the common air that we breathe. It is a fleeting Sal-nitre (Air-Salt), a mercurial-moist vapour which, when driven from the inner Central-Fire, ascends from the Earth. This moist vapour mixes in the upper regions with a dry, fiery, sulphuric mist; both join under thunder and lightning into a slimy salt-water (Chaos, seed, materia prima), which descends again to Earth, in where it coagulates to a double-vitriolic Salt or earthly Sal-nitri that is white and red in its interior (Sol and Luna).

> That is why the Hermetic Philosophers say that their Materia is being born like the thunder and leaves behind similar signs. — *(Sonne von Osten, 1783)*

The Materia of the Alchemists is threefold: the first or the remotest is called the Materia remotissima. She is a mist of the Elements, the cloud of darkness or the Water of the Depth, above which, in the beginning, the Spirit of GOD moved, the Arch-Chaos (Ur-Chaos), and it is actually the quabbalistic EN-SOPH.

The second, or the closer is the Materia remota vel secunda, in which the next or Materia proxima is hidden.

The Materia tertia ultimately encompasses the specific Beings of the three Kingdoms, consequently the Minerals, the Vegetation and the Animals.

The first Materia is the Mercurius of the Alchemists, the Universal-Spirit, the World-Seed or Primordial-Seed, and it is also called the materia prima universalis.

The Materia secunda is our subjectum crudum, our little-worldly chaos, an image of the first dark chaos. It is an Electrum minerale immaturum, the Magnesia of the Alchemists, also called Antimonium, as Basilius Valentinus attests to its meaning, as "All in All." It is a saturnine Minera, the Lead of the Sages, our Saturnus Philosophorum, a coagulated heaven's dew, a saliva of the Moon, a terra damnata, an earthly cave, the Sages vineyard and the wingless dragon.

Within this Materia secunda, the Materia proxima is contained. Our true and actual primaterial Subject, the Central-Salt of the Sages, the granum fixum, our virgin Earth, the secret Sal-nitri, our smaragdine Salt, also called Sal Saturni and Sal Tartari. It is the terra adamica, out of which the paradisiacal body of Adam was made of, which was immortal and incorruptible. The

Alchemists called this their actual and true subject, also the Root-moisture, the humidum radicale, their twofold sulphuric Salt, their Hermaphrodite or Proteus, the Chameleon of the Sages, their green Lion, untimely or premature Gold, the fixed Sulphur (outside white and inside red), their sulphuric Gold fountain, their virgin Mermaid or heavenly Nymph, in brief: their corpus solare et lunare, the greatest Secret of the whole of Alchemy.

This secret Central-Salt is called hermaphroditic because it is twofold: fleeting and fixed, male and female. The fleeting part is our Mercury, the fixed part consists of Sol and Luna, that means out of the red and the white Sulphur, they are our two tinctures (red and white). That is why it is also called the Sal rebis, the three-principled Salt, because it has within itself, all three Principles: Mercury, Sulphur and Salt.

The red Sulphur on the inside of the Salt is the Central-Fire, the subterranean Central-Sun, the Seed of the God, but it is also called the King or the Man. The Mercurius is the Water of the Metals; the Seed of the Silver, the Woman.

Even though in our virgin Earth, Fire and Water (Sulphur and Mercury) are together, they are only slightly connected with each other, so that they can separate very easily from each other, so that they lend themselves to be wonderfully reunited again.

The separation and reduction is necessary to bring the imperfect to perfection. Nature did create these Hermaphrodites, but left them to be imperfect. However, where Nature ceases to be, the Art begins, and this Art is the Alchemy.

"Because of this, it is only right that somebody makes haste with the materia and the key, until and before the whole World enters into error and the pious themselves abandon hope."
                                    — (Joh. de Monte-Snyder,
                              Tractatus de Medicina Universali).

The Subjectum Artis.

The first and crude subject of our Art is the materia remota vel secunda; the remote or second Materia. This is the Mons Philosophorum, the Vineyard of Sages, the metallic Mountain, the very best concentrated Mineral, the metallic Water, the sea or the chaos of the Sages, the tree with the three branches, the three-headed Dragon, the Cerberus infernalis triceps, the highest double Saturnus, the goldrich Crossapple, our Antimon, our Crucifer, Arbor vitae cognitionis boni et mali, the little irrational World, our Microcosm.

Paracelsus said the following about this:

"You call the Human Being Microcosmum, and this is the proper name, but you never fully understood this. You should therefore understand us and we interpret the Microcosmum. In the same way as the Heaven is constellated by itself, with all his Firmament, so is the Human Being constellated and ruled by itself. As little as the Firmament in Heaven is ruled by a Creature, is the Firmament in Human Beings being ruled by other Creatures, but it is instead, by itself a might, a free Firmament, without any commitment."   — (Volume Paramirum)

The Subjectum artis of the Alchemists is therefore, the Human Being. Not the Human Being in the common external Sense, but as an internal paracelsic Microcosm and a small worldly astral Firmament.

There are Heaven and Earth with all the constellations, the four Elements and the three Principals (Sal, Sulphur, Mercurius – that is Body, Soul, Spirit) invisibly contained inside a Human Being. We must take into our hands the first and crude Subject of our Art, to draw out Mercurius, Sulphur and Sal, one after the other. After proper preparation, out of these, the Philosopher's Stone is made.

The Primaterialistic Subject.

The true and actual Subject of the Alchemists is the inner Salt, the Sal Saturni, the Tartarus or the Winestone of the Sages, the blue-green Sal Tartari, also named the smaragdine Salt, our Vitriol, the Gold-Salt, Gold-Corpus, Sun-Body, the adamic Earth, the red Salt, the Golddust from Ophir (= Aphar min ha-Adamah), the virgin Earth, the fish Echeneis with silver and golden scales, which the Sages catch in their Sea, our Hermaphrodite, the Corpus solare et lunare, the soulful Salt, outside white and inside red, our green Lion, the fixed-green Sulphur, the Rootmoisture, the crystalline Central-Water, a fatty dry Water, the Fat of the Earth, the Mercurius coagulatus, a physical Spirit and a spiritual Body, the inner Saltbody, the Lightsalt, the Lightbody, the ethereal Astralbody.

This inner, invisible Astralbody is sometimes called Proteus; he can assume all the colours, forms and shapes of the world. He is the Corpus sideribus of Paracelsus, sometimes called Evestrum or Mumia, the microcosmic Magnet (Magnes microcosmi), because he is the life-maintaining, indestructible and attracting Principle in the Human Being. Wherever this Principle does not exist, death and putrefaction sets in.

> "In all three Kingdoms, no creature can be without this Sulphur; he is not the Life-Balsam, but virtually the glue with which the body is being kept together, and also the magnet which attracts the heavenly Sulphur or the Life-Balsam for the maintenance of the Creatures."
> — (Welling, Opus Mago-Cabbalisticum)

This Mumia, which encompasses the outer, earthly body of the Human Being like a lucid envelope, can be found in all Kingdoms of Nature, because no body could exist without this Life-Balsam's inner Magnet.

Contained within the Minerals is this Mumia, especially in the Sulphur of the same, and Paracelsus calls it Stannar or Truphat (Secret Virtue of Minerals). The plants contain this astral Magnet mainly in their ethereal oils, and here this is named Leffas by Paracelsus. In the Animal and Human Kingdoms, Paracelsus refers to this astral Organism many times as Evestrum, which has its seat in the fat and blood (hence the Fat and Blood Rites in the old Religions and Mythologies)!

The Knowledge of the invisible Human Being begins with the Evestrum, which belonged to the ancient Mysteries.

> "Evestrum is also united with the Eternal. After death, the Evestrum remains on Earth and gives indications if the Human Being experiences joy or if the Human Being is in pain." — *(Paracelsus)*

All Beings have their Evestrum, even God.

> "The most high and kind God has the Evestrum mysteriale, in which his Nature and his Attributes can be seen, and through the Evestrum mysteriale, all the good and enlightenment can be recognized. Even the one who is damned has his Evestrum in this World, in which evil and everything is recognized. That is how fragile the Laws of Nature are." — *(Paracelsus)*

Once in a while, the Alchemists referred to this Evestrum as a lasting Water:

> "The Solvens is not called Aqua permanens for nothing, the reason being that it should remain that way and it should become an Evestro." — *(Joh. de Monte-Snyder, Tractatus de Medicina Universali)*

The old Roman Philosopher Lucretius said:

> "Bis duo sunt hominis: manes, caro, spiritus, umbra; Quatuor ista loca bis duo suscipiunt, Terra tegit carnem, tumu-

lum circumvolat umbra, Orcus habet manes, spiritus astra petit."

(A Human Being has four basic parts: the manes (lit. = good, or shades of the departed), the flesh, the Spirit and the shadow. Fourfold is the place which accepts the Four. The flesh is being buried in the Earth, the shadow hovers above the grave, that is the Earth; the Soul descends to the Underworld, the Spirit ascends to Heaven).

According to the Teaching of the Ancients, Space came into being through the appearance of the Earthdisc in the middle of the hollow worldglobe, being separated into an upper and lower arch. They called the upper space the Ether or Uranos (Heaven), and the lower space the Underworld or the Tartarus. That is in accordance with the three World Teachings of the Hermetics and also with the christian Mystics.

The Earth or the Outer World is the Kingdom of the physical senses and of the common intelligence. The dark Underworld is the sphere of feelings and the chaotic pictureworld of dreams. The illuminated ultraworld, on the other hand, is the Kingdom of the Divine Spirit and the eternal primordial pictures, which are formed out of the purest Light, and reveal only infallible Truth and Wisdom.

Thus, to some extent, the Earth is an in-between Kingdom (Interregnum) between the Upper and the Lower, between the World of the Angels and the dark Kingdom of the infernal demons. The Egyptians possessed the ability of profound thinking and no one examined Nature more meticulously than they did. They depicted this in a Hieroglyph which showed the Human Being as a heavenly Creature and at the same time as an animal. This is illustrated by the form of a Sphinx, which is a Lion with a human head, with a small upright human figure standing between the paws. Between the Animal-Kingdom and the inhabitants of the Lightworld beyond, stands the Human Being, half Animal and half Angel. When the Human Being turns towards the animal, it then devours the Angel. Should the Human Being nourish (sustain) himself with the nourishment of the heavenly Spirit, the Angel conquers the Animal, and the Human Being enters into the Temple of the eternal Light.

There is probably not one picture in the whole of Nature, which depicts the position of the Human Being more clearly, shows the origin from which all Human errings are derived, or describes what Human Beings have deteriorated to and are regressing to daily.

The Sulphur of the Philosophers (Sages).

The Alchemists have three different kinds of Sulphur: an external, burnable Sulphur (Sulphur combustible), which will be separated; then there is a white, fixed non-burnable Sulphur; and then a red, fleeting (volatile) Sulphur. The last two are our tinctures white and red.

The fixed Sulphur is the green Lion in the centre of our Subjecti, the volatile is out of the bowels of our dragon extracted blood or the fire of Nature, which awakens the sleeping, in his own earthliness lying fixed Sulphur and brings the de potentia ad actum (which means, it brings it from the possibility to reality).

The green Lion is the fixed, dry, fiery Nature-Fire, that rests in the centre of the Salt of our Compositi like in a prison until it is liberated of the volatile red Sulphur, and with the same ascends towards heaven.

In the upper part of the philosophical receptacle appears, as a highly brilliant Signate-Star, our Diana, and then receives the proper Power of Transmutation, to tinge everything which is imperfect.

This astral, Central-Salt of the Earth, wherein at the same time rests our Sulphur and Mercurius, we must, out of the crude compact darkness of the Earth with the Elementwater, wherein the Upper Light is, soften, lixiviate (leach) out and extract.

Firstly, we extract a fiery Spirit out of our fixed Central-Salt, the Water of Life (aqua vitae), the volatile white Lily, the volatile white Dragon or Eagle.

In this Spiritu mercurii is lodged a subtile oil, the red volatile Gold of the Philosophers, the volatile Fire of Nature, which we extract as well, out of our Central-Salt through moderate distillation and sublimation.

This volatile, fiery Spirit will be separated from the Central-Earth, when she stands in solution, like the Light from the

Darkness, the Mercury from the Sulphur or Eve from Adam. This Spirit rises out of the Earth into the Air, and in the upper regions it will be resolved again into a fiery water, whereupon it descends again to Earth, as a golden rain or dew, incorporates with the same, moistens it and makes it fertile. Everything in the three Kingdoms has its beginning from here. The Philosophers call this moist fire the Archaeum. It is the Spirit of the **LORD** (Ruach – Aelohim), which once and even now is moving upon the face of the water, as he is also in the Air and in the Earth, where only an agreeable moisture can be found. Because he must make everything fertile through his moist fire, and temper the fixed, dry, fiery fire in the Earth. Although this everlasting Mercury always rises as smoke and enters into his chaos and disappears in the Air, the Mercury will return back to Earth, which constantly attracts him.

This, our volatile Spirit or volatile Fire, is the most refined substance of the Philosopher's Stone; the other substance is his Sal fixum incombustibile, a fatty leafy Earth, the Soil of the Philosophers, which is also drawn out of our Central-Salt.

Our Naturefire is the volatile Spirit of our Wine (Spiritus volatilis sive Spirtus vini). The Spirit of Vinegar is, however, the Fire contrary to Nature. This Spiritus aceti is not quite as volatile and therefore takes the place of the Acidum, between the Fixum and the Volatile. This sour Spirit alone is the Menstruum of the Philosophers, which has the ability to dissolve the firm corpus and at the same time coagulate himself into a firm body.

This is why the Philosophers say, that a twofold fiery Man must be nourished with a white Swan. This double-fiery Man is our moist fire or the red Mercury. The white Swan however, is the second substance of our subject, the fixed white Sulphur or the fixed pure astral part of our Central-Salt.

The Spirit cannot be coagulated, other than with the solution of the body; the body cannot be dissolved without the simultaneous Coagulation of the Body!

This is how both have an effect upon each other, and into one another; the fiery Spirit dissolves the Corpus, and this in turn binds the Spirit. Out of this emerges a kind of middle substance between the Spirit and the Body; a physical Spirit and a spiritual Body.

This is the Art-Chaos of the Philosophers, the much sought after materia prima, our Mercurius duplicatus (Mercury united with his Sulphur).

This materia prima is the pure adamic Earth, our Golddust or red Tincturepowder, out of which the new born again Spirit is now formed. This is how the old impure and ineffective Salt-Corpus is transformed into Water, and the adamic Earth is transformed again into a new, tinctorial Salt-Corpus. Because, the Stone or the Gold does not tinge, if it has not been tinged itself before: "Sol non tingit, nisi tingatur."

This is why our two Sulphurs are nothing else, than the volatile and the fixed Naturefire. The volatile fire is an external, ultra-heavenly Light, the first Radiance of GOD, the prime origin of AUR-EN-SOPH, the Light of the Infinite.

– (Gospel of John, Chapt. 1, Verse 9 – 10)

This divine Prime-Light descends out of the regions of the Ultra-Heavenly into the Heaven, that means into the Kingdom of the Constellations, where it turns into the Light of Nature (Astral-Light), into the World-Soul.

This Light preferably makes use of the Air as a carrier, through which it maintains the Life of all Creatures. While the Air condenses itself into a rain or dew, the living Fire steps secretly with the Water into the Earth, where it takes on a Salt-Body, which is a physical Water or a fixed Fire that has with the living volatile Fire the greatest inclination, because it requires the same as nourishment.

In the beginning, everything was only one Water, the prima materia universalis. Above them, the divine Spirit-Soul hovered as a dark, non-consuming fire, and that is an eternal Life. As soon as the divine primordial Fire revealed itself as the Light, through the same ensued the threefold separation of the Prima Materia into an upper, medium (middle) and lower Water.

The upper Light-Water is also named the Crystal-Heaven. It is a highly volatile fiery Water (in Hebrew, it is called Aesch-majim), wherein the influences of the eternal divine Light easily imprint themselves. This subtle Water, within our Spiritus volatilis, is therefore a duplicate of the highest of Lights, which has concealed within itself, the Soul and Life-Energies of all lower creatures.

The middle or medium Water is the Spiritus, as the carrier of the Spirit-Soul and with that the medium conjungendi animam cum corpore. This Spiritus is a living, fiery Mercurial-Water, half volatile, half fixed; a hermaphrodite Spirit which not only has an interest in the Light and the Upper Water, but also has an interest in the lower elemental Water, the Menstruum of the World. The Alchemists call this the AZOTH.

The lower Water is the rootmoisture of all things, the Central-Water of the Earth, a dissolving leafy Earth. Before the fall, this Water was pure, transparent and luminous like a crystal.

But, after the fall, it became impure, dark, coagulated Water, wherein only the uppermost, immortal energies (Spirit and Soul) are imprisoned. Through this, this terra damnata of the corruption has become subject to death, and in spite of it, it does contain pure Water invisibly within itself.

Whosoever knows how to separate the Immortal from the Mortal, the blessing from the curse, to bring the indestructible back into the original Being (as it was in the beginning) and into its original dignity, is the one, who imitates GOD'S Work of Creation in a small way, and he is a true Alchemist.

The Mercury of the Philosophers.

The Hermetics differentiate four kinds of Mercury, namely the Mercury of the Bodies, the Mercury of Nature, the Mercury of the Philosophers and the common Mercury.

The Mercury of the Bodies (Mercurius Corporeus) is the most precious among them. He is the well-cooked coagulated Rootmoisture; a certain fiery-sulphuric Rootmoisture; in short, the much sought after Lapis Philosophorum. To reach this stage is the goal, the endeavour, of the whole of Alchemy.

The Mercury of Nature is the Rootmoisture wherein the Naturefire has its mainseat. The moisture is distributed all over the whole body, even in the smallest of parts.

It is a highly subtile ethereal Corpus, a coagulated Mercury, wherein the Fiery Anima is locked up, as in a prison. When this Radicalsap receives a profusion of moisture, then it is the seed or the prima materia of the Body; if it is, however, cooked more intensely, it then becomes the Mercurius corporeus.

This Rootmoisture is the same in all creatures of the three kingdoms, with the exception that in some, they receive no cooking at all, and in others this (cooking) is done only partially.

In the Subjectum artis of the Philosophers (Sages), the Rootmoisture already received some cooking from the Naturefire or Sulphur which is enclosed within her, so that the humidum radicale already coagulated and became fixed. Nevertheless, this fixation is only in accordance to the ability, that means only potentially present, because the Rootmoisture of our Subject is enveloped with many volatile vapours (Phlegma), so that she easily evaporates and vanishes into the Air. Because, as soon as the volatile part surpasses the fixed part in a Subject, then both become volatile and leave.

Accordingly, our Sun Sulphur or Light of Nature cannot be found fixed on Earth in reality, because his Corpus (the Rootmoisture) is not yet truly fire-proof or incombustible. The fixed red Lion or Sulphur rubrum fixum is the first sought-after Lapis Philosophorum!

The Rootmoisture is also called a fatty Earth, because it is an oily Golddust, a fatty red Powder, a thick Water or Oil, this the Philosophers quite often call: the sulphuric incombustible fixed red Lilysap, or sometimes it is also called a Maccabean Fire.

The Mercury of the Philosophers is the third in our sequence. He is cleanest whiteleafed Earth, the Diana of the Sages, a peculiar Middlesubstance between Mercury and the Metals, the Sal metallorum. Out of this the Lapis Philosophorum will be immediately prepared. It is a dissolved Body and a condensed Spirit, the fixed pure astral part of our Central-Salt, that is, the fixed white Nature Sulphur, the terra foliata alba, the Sage's Sal Ammoniac Until we are in possession of this Sal fixum incombustible, this white Swan, we cannot begin with the actual Work of the Sages, the Magnum Opus.

The Mercury of the Philosophers is the most secret of all, and it is hidden by the Hermetics under many mysteries and sheaths, because it is the Mysterium magnum.

The fourth and last in our sequence is the common Mercury (Mercurius universalis). It is the volatile white Lily of the Sages, the volatile Dragon and Eagle, the Virgin Milk, the Airwater, the Dew of Heaven, the Moon-Water, the Spirit of Wine of the Philosophers, the Spiritus mercurii, the sal spirituale, the Air of the Sages, the volatile Spirit (Spiritus votalis), who has a volatile Fire (the Anima) within him. The motion of this Spirit is ascending, through which he makes all things white (pure) and volatile.

This heavenly Dew-water, ascends like a white fog out of his Central-Earth (Genesis, Chapt. 2, Verse 6), which the Alchemist must distil artificially into a fiery Mercurialwater, wherein the red volatile fire, Light or gold of the Sages is condensed to Water, and in such a manner, reaches as a moist fire, a certain heaviness which drives this red smoke, steam or Spirit back again to Earth, with which he incorporates, moistens, dissolves and makes fertile. This double-Spirit or steam is now the Spirit of Vinegar of the Philosophers, the Spiritus aceti, the red fiery Dragon, our moist fire contrary to Nature, the Menstruum of the Philosophers, the balneum mariae, Sunwater, the AZOTH of the Philosophers, the ALKAHEST. He makes the bodies red (he awakens in them the resting fire and inspires them), dissolves, and at the same time becomes coagulated.

As the Mercury has within itself an ascending into heights volatile substance, he also has one which descends and allows itself to become fixed.

The first substance is the Mercurius' own by Nature, the other however, is only potentially present in his centro; he can, however, through the Art of the Alchemist, reach effectiveness. This is why the Philosophers say, that their dissolving is of course, a natural fire, but it must be prepared by the Art.

Eugenius Philaletha said this:

"You must make this Water first, before you can find it."
— *(Lumen de Lumine)*

Lullius says this:

"Our Mercury is a Water that cannot be found on Earth, because it does not have the ability to come into effect, without the help of the intellect and the hands of labour."

Without this General-Dissolvens and the Universal-Menstruum, the Philosopher's Stone cannot come into being, especially the preparation of this Solvent (or AZOTH), which has been kept in strict secrecy by the ancient Adepts.

This is what the "FAMA MYSTICA HERMETICA" says: The solvens of the oily fiery Water, as the Universal Menstruum of all Sages, is her Secret, and the whole chemical publicum is asking for it and demanding it and it will never learn to know it, and

33

which is, without the revelation (divina aut humana), impossible to attain.

The only One, who made some comments about this Solvent was the Adept Joh. de Monte-Snyder. Although they were only hints, the ones who understood them, were the only true Hermetics. This is what he said:

> "So that I come (ad rem) to the point of my purpose, know that the Mineral and Metallic Fire on and before, is the materia prima itself, and is being found in minera Saturni, in this his Receptaculo or Universal-House; the same must depart for a while out of this Universal-House, because of fearfulness for the flying fiery Dragon, which ignites the dwelling of the cold Saturni to this degree, that he would die therein and he must surrender his Spirit; if you now can capture, this inspired Spirit in a receptacle, then you have the tali modo, I say, the Universalmenstruum, a secret Astral Fire, which has the shape of a dry and at the same time, a moist Metallic Water." – (Joh. de Monte-Snyder, Tractatus de Medicina Universali)

Therefore, a cold metallic fire is necessary for the preparation of the Universal-Menstruum, which, as Monte-Snyder says, can only be ignited by a true Philosopher. Moreover, you have to know that the Alchemists differentiate between four types of fires, which play a very important role in their secret Work.

The Four Fires of the Alchemists.

The hermetic Philosophers differentiate between four kinds of Fires: the Fire of Nature, the Fire contrary to Nature, the unnatural Fire and the elementary Fire.

The first three Fires are also called magic Elements, whereas the fourth (the elementary Fire) is only an external and coincidental Fire.

The Fire of Nature is the Central-Fire of the Philosophers, which is hidden in the Earth, and is also called the Central-Sun.

It is a Fire, solar by Nature, yet a little coarser than that (the Fire) of the heavenly Sun. This is why it is also called the Son of the Sun. He is the Archeus of Nature who moves and digests (a form of digestion = fermentation and putrefaction) the matter, when he is given his Freedom, he does all the work in her (Nature). In the interim, the Central-Fire lies hidden under a hard shell, still weak and unable. It is only a Fire in potentia, that will show its power as soon as it is awakened through the external Fire. The Alchemists call it IGNIS, or at times, call it their red Lion or Gabritius. It is the Fire of the Gold, the true Goldsulphur, the Soul of the Elements, the Light of Nature. Without its influence, the intelligence would be weak, the power of the imagination would be dead, the Spirit would be unfruitful and the body would be lifeless.

The Son of Sun deserves to be drawn out of the Darkness; to purify him and to bring him to a more mature position, for which this Gabritius still requires the help and co-operation of his Sister Beja or Diana (that is, the inner Central-Moon or Water). Because, the Central-Water, which is hidden in the Earth is lunarian in attribute, yet not as luminous, as the heavenly moon. The inner Central-Fire takes on four degrees of heat, when performing the alchemistical Work, and in accordance to how his effective attribute overcomes the other attributes

(Elements) of the Materia. These four degrees are being indicated (shown) through four main colours: black, green, white and red. All the colours which appear in the Magnum Opus come from the inner Sulphur, as the Originator and the Bringing-Forth of all colours.

Black is the colour of the Earth or the Body; green is the colour of the Rootmoisture; white is the colour of the Spirit (or Air); and red is the colour of the Soul or the Fire. In addition, the colour black is saturnine, green belongs to Venus, white is mercurial-lunarian and red is martian-solar.

The second Fire in our sequence is the Fire contrary to Nature. This Fire is, by its nature, also a natural Fire. It can only be prepared through the Art of the Alchemist. This fire is a double-hermaphrodite Fire, because it is composed of two contrary Elements, namely Sulphur and Nitrum.

Because, the Sulphur as the Central-Fire is hot and dry, and the watery Nitrum is cold and moist. Whosoever now understands how to unite these two contrary Elements with each other, has found the Universal-Menstruum, which can be compared to a unification of the cold stone snake with the fiery flying Dragon.

This hermaphrodite Spirit, this double Fire of the Philosophers is cold and warm, and moist and dry at the same time and is therefore considered to be the materia prima itself; but in this, the moist warmth preponderates the dry coldness. That is why, it is also called a moist Fire or a fiery Water. In this Fire-Water, the metals dissolve without the destruction of their Centre; out of this Menstruum Universale, the Mercurius Philosophorum is being prepared. Only this moist vaporific Fire has the capability of igniting the metallic Sulphur, the inner Central-Fire, and to multiply the Element IGNIS in our Subject.

This Universal-Menstruum is the proper Separator and the true Chymicus, who separates the impurities of the two Sulphuras. He kills and brings to life, and initiates the putrefaction and resurrection of our Subjecti, and it has the form of a Mercurius duplicatus, and is also called Spiritus aceti and is a doubly Corrosive.

He is the Spirit of Vinegar or the AZOTH of the Alchemists, the red Eagle, the ALKAHEST, and only through this double Spirit can the Soul of the King (the Gold) be brought into an Oleum.

Only this Fire-Water is capable of dissolving the Gold radically, to expose it out of its nature and to reduce it to an oily lime (calcium), or to a vitriolic Guhr, that is, to lead it back into its materia prima. Because our Son of the Sun is too deeply entangled in the Earth and bound so much so, with the Superfluity of the same, that he must be liberated from this bondage through the heavenly Vulcan, that is, through the descent to Hell of our red fiery Eagle, to deliver this our Phoenix, to redeem this imprisoned Soul out of the darkness.

Whosoever is in possession of the two magic Elements: AZOTH and IGNIS, can work out the Magnum Opus of the Alchemists; that is why the Alchemists say:

"nam ignis et Azoth Philosophorum tibi sufficunt."

The moist Fire contrary to Nature, that is our Universal-Menstruum that cannot be produced without the third unnatural Fire, which is a cold metallic Fire. This third unnatural Fire is the secret Saltfire of the Alchemists. It is a cold, dry mineral Fire, which must be transformed through a particular artifice into a moist fire, that is, into a red Spiritum, precisely into our fiery Mercurialwater. This artifice is only known to the true Hermetics and cannot be revealed publicly.

Through the steps of digestion, which follow one after the other, the moist fire will then be transformed into the natural Central-Fire and it multiplies this. It should be no secret for the seeker, who has the knowledge of the Art, that the three magic Fires of the ancient Philosophers correspond with the three Principles: Mercurius, Sulphur and Sal. Therefore, the moist Fire or the Universal-Menstruum is Spiritfire. The hot, dry, inner Central-Fire is the Metal's Sulphur and the cold mineral Saltfire is the metal's Water, which is being reduced through the Salt.

The fourth elementary Fire is a pure external and coincidental Fire. Nevertheless, it is indispensible, as far as the Magnum Opus is concerned. This Fire must not be too weak, nor too strong. In the first instance, it could not bring the inner Fire into motion and in the last instance, it would burn the matter and it would drive away the Spirit. It is the Fire of the Athanor which must be kept at all times at the same degree.

The Receptacle of the Philosophers.

The hermetic Philosophers make use of several receptacles in their work, namely those of the Art and those of Nature.

Mainly they are three receptacles of the Art: first, the open solvent dish or a crystal mortar; second, a glass vial (Phiala); third, the Athanor (oven).

The solvent dish serves for the pulverizing and calcination of the salt. The glass vial is for the boiling of the Elixir. The glass vial must be hermetically sealed and it should only be filled with one-third of the materia, besides that, it must be placed into a container or receptacle, that is lined with ashes from oak wood, because the material should only perspire moderately.

The Athanor should be made out of baked potter's clay. According to some, it could also be made out of copper. The most beautiful gems, as it is commonly known, consist of argillaceous earth (Rubies, Sapphires, Emeralds, etc.). Those who have the understanding, can easily manufacture artificial Rubies out of argillaceous earth, which cannot be distinguished from natural rubies, by whatever means. The enormous amounts of energies, which are present in argillaceous earth can be observed, since out of it Thermit can be produced, and when ignited develops the immense heat of 3000° Celsius! Why then should the Alchemists not make their Athanor out of potter's clay?

Out of which kind of potter's clay they (the Alchemists) made their Athanor, is their secret! Clay or loam, that is namely, also the mysterious basic material, out of which, in paradise, the body of Adam was made. In the Bible, it was called "Bedellion" (a fragrant resin or gum).   – (Genesis 2, 12)

Notwithstanding, the Alchemist Siebmacher, as well as Welling, made detailed accounts in regards to this potter's clay. For example Welling states: in order to be able to make out of the true Gold the Wonder-Gem Onyx, the proper Bedellion has to be added to the Gold. On the other hand, Gold is the red Tincture of the Alchemists, which in Genesis is being designated as "Adam" (Redness), that is, the Mars of the Philosophers. Bedellion, however, is Aphar or Ophir, a dust or loam from Eden, the Venus of the Alchemists. And out of the union of Adam and Aphar (Eve) comes into being: "Aphar min ha-Adamah" (red tinctorial Gold dust), the materia prima of the Human Being.

Copper is a red Venus-Metal, which is also red Earth. That is why it is also called premature Gold. Therefore, when the Philosophers speak of their "Gold" as the Corpus that should be dissolved, they are speaking of copper that must become Vitriol; that means, to become the verdigris of the Philosophers (consider the Quabbalism: Cupros = Corpus)!

In comparison, the receptacle of Nature is only one step, in all degrees of the work. This receptacle of Nature of the Philosophers is their Water, whereby it should be known that the Alchemists know an upper, middle and lower water. That means, a subtile, a moist and a viscous, dry water. It should be clear however, that the upper waters themselves must be contained and engaged in a receptacle; that is why the Alchemists say: "Aquam ipsam vase quodam contineri, necesse est omnes fateantur, et hic rei cardo est" ("All must acknowledge, that is necessary, that the Water itself be kept in a receptacle and upon this hinges the whole thing").

The Alchemists said nothing or very little about the actual hermetic receptacle that must endure all the work in the Magnum Opus. Details about this were given by the Alchemists to their Students only from the mouth to the ear. Maria, the Prophetess, points this receptacle out as being Divine, and that through God's Wisdom, it has been hidden from the Nations. Thomas Vaughan said this:

"One thing the Philosophers left out and that is the Vas naturae, viride Saturni, Hermetis. This is a menstrual substance, the Mother of Nature, therein you must place the common seed, as soon as it emerges out of the body.

The Warmth of this Mother is sulfuric, and that is what coagulates the seed; but the vulgar fire, it does not matter what kind it is, will not do it. This Materia is the Life of the Seed, because it maintains the seed and makes it alive. Without this, his Mother, he will become cold and die, and nothing real can be produced from him. Without this, his own Mother, you will never be able to coagulate the Materia of the Seed or bring it to a mineral complexion. In all of this, attention must also be paid to a certain mass (substance); without this, you will err in the Practice of this natural receptacle.

The whole secret consists of the measure and government (regimine) of this thing; that is called her receptacle, and at times, also her fire. This thing is invisible. Due to a certain reverence, it is removed from the eye, and if it comes into the approximate vicinity of somebody, it will retreat to the side in her natural manner, because it is the Secret of Nature, and that is what the Philosophers call primus concubitus." – *(Thomas Vaughan, Aula Lucis).*

The Alchemists also call this secret receptacle of Nature the Seal of Hermes, because it is the golden Magnet who attracts the sun rays and at the same time, concentrates them and seals them within him. Nothing more can be said about this.

41

The Erring Ways of the Pseudo-Alchemists.

"Whosoever, has no knowledge of the rightful beginning, will never find the desired ending; and whosoever, does not know, what he seeks, does also not know, what he will find." — *(Axiomata, Anno 1736).*

In this chapter, we will discuss the main errors of the false Alchemists, to avoid the sincere Seekers from expending unnecessary time, expense or experiencing severe disappointments. It is sometimes hard to believe how some Pseudo-Alchemists in the present attempt to solve the problem of producing the Philosopher's Stone, with highly foolish methods.

First of all, there is the Guild of so-called Air fishers, that is, Od-Collectors, who "have the need, to be self-important beyond their abilities." In other words, plusquam-perfect (more than perfect), and this is done with all kinds of manipulation (i.e. breathing exercises, mirrors, hand gesticulation).

There are those, who want to attract the Od (Life's Vitality) out of the Air, to practice some kind of sorcery with it. What the true Hermetics think about that was explained by Hermann Fictuld, who said:

"There are many who seek to catch Heaven's Salt or Sal coeleste through magnets, but we rather remain with the Minera, it is already caught in there and it has been placed there by God."

The ancient Alchemists unanimously taught that the metal seed should not be searched for in the four Elements, because these are the most remote Materia (materia remotissima), which the Philosophers do not make use of in their work.

Another group of Pseudo-Alchemists are the Metal-Toilers, who rummage around in the mineral Kingdom and they get at the ores and metals with corrosive acids and welding torches, being of the insane opinion that Gold will be obtained in this manner. It is unbelievable how much "alchemical rubbish" is offered in books to the readers for a lot of money!

Anybody who has a smidgen of knowledge in regards to the Natural Sciences knows that the basic chemical material (to which Gold, Silver, Quicksilver, etc. belong) possess a very specific atomic structure, and that the transmutation of a basic substance into another is only possible through the change of this atomic structure. This has been accomplished in a few instances through the smashing of atoms, but not one gram of Gold has been produced out of an ignoble material! Even with the most modern equipment for nuclear fusion, to produce the tiniest amount of Gold would take approximately 30,000 years! Surely, Nature did not create the basic chemical materials through smashing or disintegration of atoms. Nature creates through atom construction (that means, the atom-synthesis), however, the natural Science of today is still far, far away from this.

Besides that, the true Alchemists taught, that it is considerably more difficult to destroy, than it is to make Gold: "Facilius est Aurum construere, quam destruere." In this connection, you also have to know that the Hermetics did not operate with dead metals; they worked with live metals found above ground and not below.

The Vegetablists, that means the Plant and Herb Philosophers also had little understanding of the Writings of Philosophers, just as the Mineralists did, if they presume that the substance of the great Elixir is contained in the sap of plants. Although the Alchemists speak of a vegetational Root-Moisture, but, which is also mineral, animal and astral, that is truly universal. The true Hermetics pointed out, again and again, that the substance of the Philosopher's Stone cannot be found in the three kingdoms of Nature (that is, in the Mineral, Plant and Animal Kingdoms), since the creatures of these three Kingdoms already belong to the materia tertia, that is, to an already specified materia.

43

The Animalists are barking up the wrong tree as well when they believe that the Materia of the Philosopher's Stone can be found in animal substances (such as Blood, Body Substances or even in the excrements). These Stercorists mess about just as the Vegetablists and the Metal-Toilers in the materia tertia, without any thought that the true Alchemists took only the unspecified materia secunda as their primary material!

Lately, a new category of Pseudo-Alchemists have emerged, who seem to think that they can produce the lapis Philosphorum with the aid of a certain occult psycho-technique. This group could be called Illusionists, because they fool themselves and others with imaginary hokuspokus.

Since this group refers to the testimony of ancient mystic authors in many instances, it has become necessary to respond to this and take a closer look at their peculiar eccentricities.

The direction that the pseudo-alchemistical Illusionist is taking, comes from the so-called "Sebottendorf Exercises," which a certain Schwidtal supposedly brought back from the near East (Orient), which Sebottendorf published under the title: "The Practice of the Ancient Turkish Freemasons."

It concerns certain intellectual concentration exercises coupled with specific hand gesticulation (Mudras). Therefore, the practitioner, for example, has to imagine that the extended index finger of his right hand represents the letter "I"; when the hand forms an angle, the letter "A"; and the bent thumb and index finger, the letter "O."

When these exercises are being done with a certain amount of perseverance, the particular practitioner should firstly experience a Warmth in his index finger, then perceive a sulphur odour, and also the taste of salt and quicksilver sublimate. With this, the three main principles of the Alchemists, namely Sal, Sulphur and Mercurius has been mentally "acquired" by the Illusionists! Furthermore, when doing these exercises, the "Exercitand" has to concentrate internally upon the syllables Si, Sa, So, then upon the syllables Alam, Alamas, and Alamar. He then must hone his "Inner Sight"; then before his spiritual eye appears a "Black Shadow," the so-called "Raven's Head" of the Alchemists or the "materia cruda"!

When these Si-Sa-So exercises are being continued, then the black shadow becomes grey little by little, blue, scarlet red, pale green and eventually deep green. With this, the practitioner acquires the "Peacock-tail" (= train) of the Alchemists! Then the imagined "Shadow" becomes, through concentration upon the formulas, ALAM CHAM, CHAM, etc., brilliantly white. With this the Illusionist believes to have produced the "White Swan" of the hermetic Philosophers. Eventually, through further Letter-Exercises, the "materia cruda" becomes magnificently red. The work is finished, the stone of the Illusionist has been reached.

It must be clear to anybody who has an understanding of this subject, that this has nothing to do with Alchemy.

A similar system, like the Sebottendorf Exercises, was taught by the author Weinfurter from Prague in his Book "The Burning Bush," although he did not recommend the concentration upon certain finger gesticulations, but instead, to do the letter-exercise in the feet.

Weinfurter referred mostly to the Mystic J.B. Kerning, and it is quite apparent that he did not understand him totally. In my opinion, the Letter-Exercises of Sebottendorf/Weinfurter are nothing else but a form of Self-Hypnosis, where in the course of this, they turn into hallucinations.

Whosoever practices such self-hypnotic exercises with corresponding perseverance and without having true knowledge, will eventually not be able to get rid of the Spirits, as in the case of Prof. Staudenmeier, who thoughtlessly called upon them. This path can lead to Schizophrenia, if not worse, namely black Magic.

The theosophical author Franz Hartmann wrote about this:

"In Germany, the exercises which are described by J.B. Kerning are often in use, which consist of the repetition of certain words, in the mind. At the right place and in the right manner, these exercises are without a doubt of benefit; but, since it is many times the case, that the teacher, as well, as the student, have a totally wrong idea about the exercises, and believe that through thoughtless reciting of formulas and spells they come into the possession of magic powers so that they can use them for selfish purposes, this path leads to ailments and insanity."

– (Dr. Franz Hartmann, Geheimschulen der Magie)

The Mystic J.B. Kerning made use of these Word-Exercises only as an emotional stimulus. Kerning was of the opinion that a Human Being, in order to reach the level of free speech, had to first learn the primary teachings of the language, namely to learn to think and to perceive the letters, then the formation of the words = morphology, and eventually to Syntax, to form sentences.

Kerning says verbatim:

"We should try, instead of inhaling unseasoned air, to inhale vowels, then words and finally whole sentences, and then allow them to penetrate the whole body. You will soon experience, what it means, to expose yourself to coincidences or to nourish yourself with the Air of Life." (Kerning – Die Missionàre). Kerning did not think it was advisable to practice only vowels, as Sebottendorf and Weinfurter predominately taught: "To practice only with vowels is a difficult task, since their currents do not reach a standstill and therefore it is very difficult to feel them."

– (Kerning, Briefe über die königliche Kunst)

The only thing Kerning was trying to obtain with this method was that the practitioner would not recite certain sacral words and prayer-formulas purely intellectually, but they should also be perceived by the Soul. His basis for this was the quabbalistic concept that the WORD was the arch-principle of Creation. Each Letter of this Arch-word is therefore an Element of Creation (stoicheion). The whole of creation is God's perpetual Speaking. This Speech (or Language) manifests itself in the forms and shapes of Nature and a Human Being should learn to perceive within themselves. Each and every thing, energy and appearance in Nature has its own character, and these characters are God's powers of speech, which Human Beings should learn to recognize and perceive as Letters of the divine Word of Creation. The Spirit creates by movement. The movements have a form, a character, a peculiarity, a direction, a mission. All movements can be considered to be variations of the threefold fundamental theme, namely straight lines, angles and circles. Since the latin-greek letters I A O have, in the manner in which they are written, a certain similarity with a straight line, an angle and a circle, it was Kerning's conclusion that vowels I A O must be the roots for God's Name, but this does not apply in the Hebrew language or in any other language. In Hebrew, writing the letter "Jod" indicates the vowel "I" as well, as the vowel "E." Aleph is very seldom used as a vowel, and for the vowels "O" and "U," there are no letters in the Hebrew language, but only the consonant WAW. Besides, the Hebrew Jod, Aleph and WAW do not have the slightest similarity with a straight line, an angle or a circle, as Kerning thought it to be. Also, in all the other semitic writings, single letters of vowels are almost unknown; that is why it appears to be very doubtful that of all people, the Arabs should have invented the I A O-Exercises, as Sebottendorf declares.

Why then, of all things, should the WORD be the arch-principle of creation, as the Quabbalah teaches? Why not, just as well, the NUMBER, as Pythagoras expounded? Or, the geometrical Principle, as proclaimed by Plato? It could also be said, that everything was created through the LIGHT, and that the whole of Nature in all its graduations, represents only a colorful scale of the Primordial-Light. Without Light, there is no Life! It does not matter if (it is) WORD, NUMBER, LIGHT or the geometrical Principle; basically, all of this is One, namely an expression of the activity of the eternal divine Wisdom. We must not mistake the eternal WORD of GOD with the temporary words of Human Beings, and we should not mistake the divine PRIMORDIAL-NUMBER with the numbers of the Nations' Shopkeepers, and the world-creating PRIMORDIAL-LIGHT, with the Colour-Hallucinations of the hypnotizing Illusionists!

The human language serves only to communicate to others, in words, expression of feelings and thoughts, for the purpose of reciprocal understanding. The original languages of the primitive nations, as well as the higher animals consist mostly of sounds of perception (vowels), and the following effects are giving an expression (i.e. joy, contentment, discontentment, pain, annoyance, anger, etc.). These are pure concrete languages of feelings or emotions. With the unfolding of the intellect, the Language of Conception, with its abstract word-formations, came into being, little by little among Human Beings.

A language is not only a communication to the outside, not only an utterance, but the language can also become a speaking to the inside; it can become an inherent awareness. The Human Being is internally divided into a conscious I-PERSON and into an unconscious DEPTH-PERSON, that means Spirit and Soul. The Soul becomes a littler clearer in a dream, when the conscious "I" has to suffer without being able to resist, may these thoughts or activities be beautiful or ugly, true or false. The Soul is the mirror of the Human Beings, that means = the conscious "I," in which the Human Being sees his thoughts and the Life of the Senses personified and taking place. In Truth, the Soul is the immortal Spark of God in a Human Being, his given-higher "I" from God. However, this spark is enveloped with the darkness of the earthly materia and lies powerless in the prison of the body.

48

It is now up to the Human Beings to awake his Inner-Life, and to bring this, his divine Soul to speak.

With his Word-Exercises, Kerning wanted to achieve, that his students would learn how to speak to their innermost and that was the purpose and nothing else. Kerning presumed that the power in the quabbalistic word would lie only in the composition of the letters.

> He was of the opinion that the Egyptians, Indians, Chinese, Persians, Israelites and to a part the old churchfathers had given their believers aphorisms and prayers, which they were to speak to their innermost as letter-exercises, without any regard as to their content. Kerning thought that the Authors paid more attention to the manifold positions of the vowels and consonants in these prayers and aphorisms, than to the conceptual content.
>
> – *(Kerning, Briefe über die königliche Kunst)*

There is only one thing that can be said about this: Here (in this instance) Kerning was in error! It is obvious that Kerning was influenced by the Quabbalist Isaak Lurja, who taught a system of the Letters-Kawwanot. However, what the proponents of Chassidim themselves think about this system comes plainly into view, out of an account which Martin Buber gives, and here is his quote:

> "When in the days of Rabbi Pinchas from Korez, the Prayer-Book which was totally based on the Letter-Kawwanot, which carries the name of the great Quabbalist Rabbi Jizchak Lurja was published, the students of the Zaddik asked him for permission to pray out of this prayer-book. However, after a little while, they approached him again and they complained, that since they had been praying out of this book, that they suffered a great loss of the feeling of the vigorous life in their prayers. Rabbi

49

Pinchas answered them as follows: "You placed all your energy and purpose of determination of your thoughts into the Kawwanot of the holy Names and the Letter-Entanglements and you diverged from the fundamental importance: to make the Heart whole and devote it to GOD. That is why you lost the Life of Holiness and the sense of feeling."

— *(Martin Buber, Erzàhlungen der Chassidim)*

This very clearly explains, of what value these normative, that means purely formal prayers, are, and how detrimental the consequences are upon the inner life. Let alone, the Vocal-Stammering of Sebottendorf-Weinfurter, which are a far cry behind the prayers of Lurjas.

In Antiquity, the Belief in the magic Powers of the *god-given* Names and *nomina sacra* were widespread. In those days, the "NAME" of a thing or a being had a much deeper and livelier significance than today. It was, in a manner of speaking, the character of a person or object itself, in a way, their mysterious representative or Doppelgänger (Double).

Giving a "NAME" was a magic operation, because within a name was the expression of the designation, the target, the reason for being. Therefore, with a change of name, the character and the nature of an object or person changed. For this reason, the Neophyte was given a new name when he was admitted into a religious order, which usually was connected with baptizing. This was not only indicative of being symbolic in regards to a fundamental transformation of the character and a conversion into a new psychic-spiritual existence, but it was also nothing short of its magical effect. With the name, the character of the Neophyte changed.

The Neophyte of the ancient Mysteries also received, during the festive Initiation, the Name of the Deity, which he had to keep strictly confidential. Through receiving this Name of the Deity, the Mystic himself became the carrier of the nature of the mystical Deity, therefore sharing her supernatural powers and mercy.

Since the Deity is in possession of a wealth of magical divine powers, she is therefore, also in possession of a great number of names. These mystical divine names do not consist of senseless compositions of letters, but they indicate exactly each and every aspect of the nature of the Deity, and connected to this is the divine emanation of power.

According to the Bible, Moses was given such a miraculous god-given Name.   – (Exodus 6. 2, 3)

The disciples of the first christian congregation were handed down such a magic, effective spiritual name. Even outsiders, who somehow came into possession of these names, would experience the same success, if they would make use of such a name. Jesus himself made use of certain *nomina sacra,* when He engaged in magic-therapeutic activities, just as his disciples did.

– (Matthew 8-16 / Lukas 10-17)

The whole 17th chapter of the Gospel of St. John deals with the communication of the mysterious *god-given* Name to the disciples.

Having received such a spiritual name represents a high and holy obligation, because through this name, the Human Being should be reminded of his true intended purpose and his higher mission. This name should be his ally, his energy in opposing his cravings for doubt. This Name should raise the Human Being out of the animalistic and material world and should give him the consciousness that he is carrying God's Power with him.

– (St. John 1-12)

It is evident that even Kerning subsequently changed from his initial teachings, the primitive exercises of senseless, meaningless compositions of letters, and he then turned to the well-defined, meaningful *nomina sacra.* He recommended this to his students, not only as the means to change the mood, but as a direct key to transmute the character and for spiritual rebirth.

With respect to this, Kerning writes the following:

"The external Human Being cannot obtain with his knowledge and his activities, positive immortality. It is therefore absolutely necessary that in his Innermost, in all the organs of his body, his "I" speaks his Name and that he can perceive these organs, speaking again, to him.

When the Soul and the Spirit remain in our Flesh and Blood, and they do not become alive to a free "I," then they are still buried in the Earth, and when the corruption occurs, they must out of necessity, dissolve into atoms. However, as soon as we deliver them from their imprisonment, and we get them to speak, they incorporate the outer "I," then they also become Lord and Ruler over the earthly life, and they will lead us, since we became one with them, in Triumph to Immortality. It is therefore imperative and our duty to have our Soul speak, so that we can have a dialogue with her and receive from her the infallible Laws for our real reason for being."

– *(Kerning, Briefe über die königliche Kunst)*

And Kerning continues:

"Each and every Brah'min, who has reached a certain level, receives from the higher Brah'min his "own Word," which must serve him as a means of disposition in all matters of life i.e. when speaking, prayers, helping and comforting the ill and those in need. The first christian Churchfathers made use of these means with great success. But this is dependent upon learning and feeling such a word within yourself, then its effect is certain and infallible, like no other word in the whole of creation.... The manner, to tune yourself with words, is the final stage of thinking, the Crown of Freedom, the inextinguishable Light of the Philosophers, and it is the often falsely understood Philosopher's Stone, who has made out of many unwise, fools; it is the key to the true recognition of God."

– *(Kerning, Die Missionàre)*.

When it comes to such means of tuning or means of changing feelings, only certain words or names can be considered and not just any senseless combination of letters that has been proven by Kerning through many examples in his writings.

At this point, we will mention another statement Kerning made in regards to this subject, to clearly show that you can find two totally different Systems of Teaching in his writings. Kerning also said this:

> "Whosoever is not afraid to make the effort and possesses perseverance, will come to the conclusion as to how far-distant the Human Being is from his spiritual "I," and what he will gain when finally, in the Innermost of his Heart, he will again learn how to call himself.
>
> The greatest part and focal point of these instructions is regarding this "Inner Calling." It is, however, after the bark has been broken and divided into several steps, because Human Beings consist of many different energies, but they, each and every time, give the "I" another name to elevate it to his highest honour." – (*Kerning, Wege zur Unsterblichkeit*)

It was Kerning's intention, as aforementioned, to lead his students to the point where they not only think certain sacral words spiritually, but they should learn to feel them with their whole heart as well. That means, astrally; yes, even perceive them through the whole body.

In this connection, Kerning conceived the Human Being (the external appearance) as a ladder, a gradation which the student should climb spiritually, from the bottom to the top, by means of the letter-exercises; namely, the Freemason-Apprentice is in three steps, the Journeyman is in five steps, and the Master is in seven steps.

This is probably one of the oldest, somewhat misleading masonic concepts of Kerning. According to this, it seemed possible to him, in this manner, to learn to perceive particular words on a purely physical basis.

In regards to this speculation, Kerning went as far as dividing each and every step of the seven steps of the Master into seven additional steps. On account of this, he complicated the Human Form into 50 divisions and even more. Then, all of them had to be dealt with accordingly, "letter by letter"; that means, to vitalize them to the degree that they become able to perceive.

It seems however, obvious, when Kerning drafted this speculation, which he considered to be free-masonic, that it became apparent to him, that it was uncanny. At the end of his elaborations, he wrote with unmistakable skepticism:

"I do not know how you will get out of the labyrinth, into which I led you."

The last sentence allows you to look deeply into this matter, and it cautions you not to make haste and to think first, to avoid taking any unwise steps in this direction. It is fairly easy to lead someone into a labyrinth, but it is considerably more difficult to guide them out, safe and sound.

Even though you will find such profound truth in certain parts of Kerning's Writings, that does not mean you should not be very skeptical when it comes to some of his statements, especially those which deal strictly with the Letter-Exercises; the latter are the ones which Weinfurter constantly refers to. Here he gives proof that he imitates Kerning thoughtlessly; he copied the writings of Kerning, of which Kerning himself was very uncertain and most problematic.

There is not one person known to us that became a better Human Being, or was spiritually reborn through such psycho-technic hokuspokus; you shall recognize them by their fruit!

It seems that the followers of the psychotechnic Letter-Exercises have not come to the realization that the **Word of God** should not only be contemplated and perceived, but most of all, it should be learned. Be **Doers** of the Word, not Listeners alone! With GOD, we shall do mighty deeds; that is what is written in *Psalm 60 – 14*. When the disciples asked Jesus how they should pray, He explained that it is not important how many words are said, because God knows beforehand what the Human Being is in need of. Before they call, I will answer!

– (Isaiah 65 – 24 / Matthew 6 – 8)

The Lord's Prayer *(Matthew 6 – 9 – 13),* which Jesus Christ taught his disciples, is therefore, not a Meditation-Formula, nor is it a Petition-Prayer, as is the opinion of most people. In reality, they are instructions of how, in practice, to deal with things!

The first sentence of the Lord's Prayer: *"Our Father which art in Heaven,"* already demands of us, that the inner attitude of our Spirit should be upon the word "OUR." It does not say *"My Father"* but *"OUR Father."* Most of all, we should awaken within us, the consciousness of a universal union with GOD that unites us in Love and God-son-ship with all beings.

"Hallowed be thy name." Through us, the Name of God should be sanctified by accepting His name in us and allowing it to become flesh and blood in us, until CHRIST acquires a Form or Shape in us! *– (Galatians 4 – 19)*

"Thy Kingdom come." The Kingdom of GOD does not come in an external manner; it must come into being internally within us, once we have removed all the obstacles in us which, up to now, prevented the COMING of GOD'S KINGDOM. The Kingdom of God is righteousness, peace, and joy in the Holy Spirit.

– (Romans 14 – 17)

"Thy Will be done on Earth as it is in Heaven." Through us, God's Will should be done, but not by calling "LORD, LORD!," but by giving up all of our bad characteristics and do the Will of the heavenly Father! *– (Matthew 7 – 21)*

"Give us this day, our daily bread." Here, the spiritual bread is meant, the "BREAD of LIFE" *(St. John 6-48),* but only those who hunger for it, will get it. *– (Matthew 5-6)*

"And forgive us our debts, as we forgive our debtors." We cannot receive forgiveness from God for our debts, if we do not first forgive those who owe a debt to us. *– (Matthew 6-14,15)*

"And lead us not into temptation." This means that GOD may

protect us from prematurely testing the magic powers which have awakened within. Therefore, this is a directive which requests that we resist the demon and draw nearer to GOD.

– (James 4-7)

"But, deliver us from evil." The greatest evil is our own lower Nature, the evil within us, that we should overcome through the Power of God.

This makes it obvious, that the Lord's Prayer is a directive for our practical course of action, and it does not represent a psycho-technical "Paternoster-Persuasion-Exercise," as a Pseudo-Rosicrucian recently tried to convince his readers of. It is also totally absurd, when the followers of Sebottendorf contend that Jesus Christ taught his disciples such psychotechnical hokuspokus.

It is unmistakably clear, when it comes to the Gospels and to the Epistles of the Apostles of the New Testament, that Jesus Christ first and foremost taught the vita activa, and not the vita contemplative!   *– (Matthew 7-21 / Corinth. 13-1, 2)*

The higher value of the active Life, compared to the purely contemplative, becomes clear only when you think about the following:

According to Paul, a Human Being consists out of the trinity of Spirit, Soul and Body.   *– (1. Thessal. 5-23)*

The Spirit is the Judgment and the Will-Principle within us; the Soul is the Feeling and the Conception-Principle; and the Body is the implementing or the acting-principle, the Executive-Organ. One without the other two is powerless.

The Soul, being the inner sensorium, lives in her own picture world. These pictures are collected by the Spirit through abstraction into concepts, to draw judgments out of the conclusion; that means, to comprehend the sense and meaning of the spiritual ideas and primordial images of the pictures seen.

When the Spirit reaches an agreeable judgment, he then brings, through the Power of his Will, that particular idea, through the physical action into realization. The Spirit being the

Will-Principle, he alone has control over the motor Nerve System of the body, upon which the Soul, under normal circumstances, has no influence.

In turn, however, the Soul controls the sensitive Nerve System that conveys external and internal observations and perceptions, which are withdrawn from the influence of the Spirit (the Will).

The Spirit is therefore the medium conjungendi animam cum corpore.

As soon as the Spirit bestows his consent to a concept or an astral ideal (wishful thinking), then out of this condition of mere potentiality, which it had in the Soul, it changes into a state of true reality, that means, it becomes flesh and blood, it becomes substantial, personifies!

All of the powers of the Soul strive for incarnation, for substantiation. Without the body, the Soul is powerless, without the Spirit will-less.

The more a Human Being occupies himself with an idea, and incorporates the idea within himself, and realizes it through doing, the more the idea becomes substantial in him. This, so to speak, in him, flesh becoming an idea, forms a new second "I," a new spiritual Being, a kind of rebirth in a Human Being.

This new "I" is not immediately complete. It requires a considerable amount of effort and perseverance for its formation.

All ideas that the Human Being accepts are spiritual Seeds and pro-creative energies. The spiritual idea is the seed, the Soul is the soil, and the body (the astral body or astral form) the fruit. Whatever the Human Being sows is what he will harvest. The Sage produces a heavenly rebirth, the fool and the villain produce a hellish rebirth within themselves. Whenever the physical body falls apart, then the inner Life stands there, in the form of the Astral body as a product of the imagination; may it be in the form of a pure light angel or in the form of a dark, angry, fiery demon.

The righteous will then be illuminated like the sun in GOD'S

Kingdom. However, should the Soul be darkened, the whole Astral-body will be dark.   — *(Matthew 6-23 and 13-43)*

At this point, whosoever is in possession of some intuition, will have an idea as to the true meaning of the alchemistical colours, which have nothing to do with the auto-hypnotic hallucinations. They are a symbolic expression of inner-psychic conditions and emotional forms. The Soul speaks in our heart through feelings which preside therein; the Soul itself is the feeling. Her shape is the form of feeling of the Spiritual. Out of the Spirit (thought-light) come the Light-Forms (Ideas), and her form is called "Word." The Soul is the envelope of the thoughts. Thoughts can only unfold into divine feelings in the Soul, because the Soul is the reflection of the Divine and the immortal Life in a Human Being. The Body (Astral Body) is the active form of the Soul; it is the incarnation of the word or thought, it is the coagulated Spirit. Only through the Soul, to whose pictures he owes his ideas, can the Spirit coagulate into a body.

Whoever thinks about this with the deepest of understanding, will know why the coagulation of the Mercurii, with the help of our philosophical Martis can be found in Saturno, as the Alchemists said!

A Human Being does not have to speak to GOD with external Words; GOD already knows what he wants. Instead, Human Beings should let GOD speak within them, they should accept within them the Word or the Name of GOD, firstly through the thought, as spiritual recognition, then through the Soul, as a feeling and finally through deeds, we should attempt to act from the inside to the outside, and not from the outside to the inside!

The so-called "Persuasion-Exercises" of mindless vowel and syllable compositions are senseless and false. Will-o'-the-wisp (Ignis fatuus) are those who pretend to be Teachers of hermetic things, yet do not have their teachings from GOD, but instead, through false or misunderstood books, hearsay or external intelligence.

The Process of Magnum Opus.

"Here I give you my key and My seal;
one opens, one locks: Use both with understanding."
— *(Eug. Philaletha, Lumen de Lumine)*

Although some Researchers have, to a certain degree, the theoretical knowledge of the Primary Material of Alchemy, which is the inner Salt-Body, there is, in spite of this, a considerable distance that has to be covered until the practical preparation of the Philosopher's Stone can be reached. The Materia is relatively easy to find, but the true Praxis of the Magnun Opus is much more difficult, since this secret process is not openly explained by any of the hermetic Philosophers.

The Adepts say that they discovered everything except the preparation, and since this is the Greatest Secret of the Art, no one else but GOD can reveal it. He will give it only to those who deserve it!

The Son of Sendivog said:

"There are many, who are under the impression, that they have the knowledge of the preparation of the philosophical Saturn, but after they have been tested through our Red Squire, it is hard to believe, how few there were – how small their number, who passed this test. Where do we find such a book, which gives us enough information in regards to this matter? Since the Philosophers practice silence when it comes to this point, and they want this point kept hidden and concealed."   — *(Compass d. Weisen, p. 283)*

In conversation with Pyrophilus regarding the War of the Knights, Eudoxus says the same:

"Many have the Stone in their possession; some despise it, as a valueless thing; others are in awe of it, because of its supernatural signs in its birth. Some have the knowledge, that the Stone is the true subject, but they do not know its true preparation, which is so concealed, as well as the Stone in the Philosopher Mercury, as well as changing this step by step into the tincture."

— *(Exegesis of the War of the Knights)*

It should, therefore, be no surprise when the uninitiated scribes, who do not understand the Writings of the ancient Adepts, again and again dream up the most adventurous processes. They are trying to impress their readers, who are mostly no more informed, than they are. To put it more bluntly: to lead them astray.

Some time ago, a so-called "Rosicrucian Expert" considered the Formula of the Arcanum to be, which was given by Michael Maier in his "Themis Aurea" as an Incantation Exercise à la Sebottendorf!

The Adept Michael Maier wrote in his 1618 "Themis Aurea"

"The knowledge of the Arcanum should be the key. I will give you the secret: t. tmmb. id. i. a. ofi. c. qqc. x Open if you can!."

For the Knowledgeable, the solution of this formula is clear and understandable: "The true Mumial body is drawn internally, artfully, out from its coarse quasiquidam cover ! x."

X is the Denarius, that means, the true Arcanum:

"SAL Antiquissimum, LAPIS, Mysterium! cujus Nucleum in DENARIO, Harpocratice SILE. (x)."

— *(Heinrich Khunrath in Amphitheatro, Fig.III, Quaest. 2)*

The Pseudo-Rosicrucian will never discover the true secret!

Everything a Human Being does with diligence and perseverance, comes alive in him and takes on a shape and form which strives to attain dominion over him; that means, to become the controlling "I." Good and evil forces can be awakened in the Soul in such a way and be brought to a sort of rebirth.

In earlier days, these forces were called spirits and demons. The Good brings a heavenly force, brings within himself a good Spirit or a Guardian Angel to a rebirth. Evil brings a hellish force, an evil spirit or devil. The hellish spirits or demons are spawns of the Devil of the wrong life's passionate desires of Human Beings; the heavenly Spirits or Angels, by comparison, are the personification of high virtues.

Such a force will not always immediately turn into a ruling "I" within the Human Being, but instead, generally into a "Part-I."

In such instances, there seem to be two or more "I" complexes present within a Human Being at the same time. These often contradict each other and quarrel amongst each other about the supremacy.

Should the Human Being be successful in strengthening the pure original human nature and elevate it to its true dignity, then little by little, "the devils" will flee from him. At this point the Human Being, who has been previously plagued by demons, emerges as a "Born Again Human Being" from this inner quarrel, and paradise opens itself up to such a person with its divine wonders.

This is actually the true Process of the Magnum Opus of the true Alchemists, which for centuries, was completely wrapped up in a mysterious darkness.

Of course, keeping this a secret, had valid reasons, because through the wrong and improper employment of the alchemistical process, the greatest calamity could come into being.

Therefore, this process is necessary, because the Soul cannot be born again in an impure body. With such a rebirth, a conscious continuous life after death cannot be thought of.

This is why Eug. Philaletha said:

"The tincture or the Soul of Sulphur cannot be new-born

in her own impure body, she must leave the earthly dark house and take on a new pure body before she can unite herself with the heavenly Light."

    – *(Eug. Philaletha, Euphrates oder die Wasser von Aufgang)*

In the subject of our art, mercury and sulphur (Spirit and Soul) are together; but they are linked so weakly, that through the art, they can easily be taken apart, cleansed or purified and in a wonderful manner be newly united again.

Without this separation, purification and reunification, the danger exists that the Spirit, when death occurs, will escape and the Soul will be without a guide and must wander around aimlessly in the dark chaos of the underworld, and in a way experience a second death, that is a condition the Bible calls "Hell."

Under certain circumstances, this condition can occur before death. This is due to the fact that once in a while a total separation of Spirit and Soul can take place during our lifetime. Through this, a Human Being falls victim to a hopeless obnubilation and becomes the booty of the Demons. These occurrences are not infrequent, the mental hospitals are full of them.

In our present times, those who should be knowledgeable in matters of the human soul (Psychologists) are mostly perplexed when they are confronted with matters like this. They do not know what the soul really is. Since they do not acknowledge a Spirit and a substantial Soul existing independently from the body. This is why they constantly confuse the concept of Spirit and Soul. For example, in one of the newer philosophical dictionaries, it is stated:

"The Spirit is synonymous with the Soul, or in contrast to the emotional and instinctive life: the "higher" Soul-life, the thinking-Soul...."

When such confusion exists, you should not be surprised over such Contradictio in adjecto. When at one point, the Spirit is being placed as the "thinking principle" in contrast to the Soul, the "instinctive life," and immediately thereafter, it is declared to be the "thinking soul" and identical with the latter.

In a Human Being, we basically have to, upon closer examination, differentiate between seven Principles:

1. According to the substance; the coarse external physical body belongs to the Mineral Kingdom.
2. The vegetative life; the Vital Principle, is that which is plant-like in the Human Being.
3. Instinct or the instinctive life; the unconscious Depth-of-the-Soul, the animality, synonymous with the animal in Human Beings.
4. The emotional and perceptional life; the picture-world of dreams and fantasy, the inner sensorium, the conscious soul is: *the child in us.*
5. The thought and judgment principle; the Spirit (intellect), the "I-Consciousness" and the personal will is: *the Human Being in us.*
6. The Principle of the higher intelligence and understanding; the geniality (Intuition) and wisdom, the higher SELF, the divine Spirit-Soul is: *the Angel in us.*
7. Above all, on a throne, God's Eternal Primary Light; the highest principle of Human Beings, usually overshadows Human Beings more or less and pours only its rays into the higher Spirit-Soul, without totally penetrating into the Human Being.

Agrippa von Nettesheim proves this in his "Secret Philosophy" when he writes:

"That is why Human Beings alone enjoy this honour, they partake in everything, work together with everything, and they are in contact with everything. They share in the materia of their own subject; when it comes to the elements through their fourfold body; when it comes to the plants through the vegetative energy; when it comes to the animals through the sensory life; when it comes to Heaven through the etheric Spirit and the influence of the upper parts upon the lower parts; when it comes to the Angels through their reason and wisdom; when it comes to God through the personification of All. Human Beings associate with God and the Intelligences through Belief and Wisdom, with Heaven and the heavenly through reason and speech and with all the lower things through the sensory life."

Nettesheim understands by reason, divine reason, the principle of understanding and wisdom. With reason, he understands the thought and judgment principle. That understanding becomes prominent in a certain part of his "Secret Philosophy," when he writes:

"Plotinus and all the Platonists accept, as did Trismegistus, the position that Human Beings are threefold, namely, the Uppermost, the Lowermost and the Middle. The Uppermost is the Divine, which is considered to be the intelligence, the higher part or the enlightened understanding. In Genesis, Moses calls this the Breath of Life – inspired by God or His Spirit. The Lowermost is the sensitive Soul, which is called the picture (Image). The Apostle Paul calls it the animalistic Human Being."

The Middle is the reasonable Spirit, which is the extreme of both, and connects the animalistic Soul and the Divine Intellect and takes part in the Nature of these two extremes. Yet, the reasonable Spirit is different from the Uppermost in Human Beings, that is the illuminated understanding, the Divine Intellect, the Light and the Higher Part. It is also different from the animalistic Soul, from which we have to separate the reasonable Spirit, through the power of God's Word. The Apostle Paul describes this as more penetrating than a double-edged sword and of which he says that it is alive, effective, and that it is capable of separating the Spirit from the Soul.

> The Higher Part in us never sins, never consents to evil, always resists error and only gives counsel for the best. In this manner, the lower part submerges itself, the animalistic Soul, always into evil, sin and carnal desires. This is the Law of the Limbs, of which Paul said, it imprisons us in the Law of Sin. The Upper Part in us will never be damned, but instead returns to its origin, untouched by the punishments of its companions. The Spirit, which is called the reasonable Soul by Plotinus; however, according to its nature is free, can in its own discretion attach itself to both parts, but will, when he consistently attaches himself to the Upper Part, unite with the Upper Part and is transfigured into a blissful life, until he is accepted by God. However, should he attach himself to the lower Soul, he then becomes subject to sin and becomes worse and worse, until eventually he becomes an evil Demon."
>
> – (Agrippa von Nettesheim,
> Geheime Philosophie, Bd. III.)

That is exactly what has been said in regards to the separation of the Spirit from the Soul. If the Spirit of a Human Being is not being illuminated by higher understanding, then he is not free from error. The divine intelligence of a Human Being does not lend itself to provide the Spirit with Light, unless it is being illuminated by God Himself, the Prime Light.

The Divine Intelligence in Human Beings is above fate and rests upon Divine Providence. The Soul however, is subject to the Might of Destiny. In the meantime, the Soul totally retreats into the Spirit, where she either thinks about other things or observes herself (the State of Contemplation).

Through the Spirit, the Soul can ascend into the Upper Intelligence, where she is being satisfied or fulfilled by the Divine Light (the State of Contemplation).

In the language of the ancient Mystics, a Soul, which is united with the Divine Intelligence, is called a standing Soul. Such a Soul is truly immortal. A Soul which is submerged in the lower and the animalistic, is called a fallen Soul, and will not reach immortality, but will instead, perish with the animalistic part, as is written in the Scriptures   – (Ecclesiastes 3, 19):

> "For that which befalleth the sons of men befalleth beasts;
> even one thing befalleth them: as the one dieth, so dieth
> the other; yea, they have all one breath; so that a man has
> no pre-eminence above a beast, for all its vanity."

It should now be understood, as Geber states in his writings in respect to Alchemy, that nobody will ever reach perfection in this Art, who has not become cognizant of these principles within himself.

Everything is therefore dependant upon the Spirit, the Middle Principle, between the Divine and the animalistic in the Human Being.

It is therefore of the utmost importance, when it comes to the Magnum Opus of the true Hermetics, that the Soul and Spirit join into an inseparable union. What is meant by this, is the union with the Divine Spirit, because only through this can the Soul of the Human Being reach immortality.

The Soul is, as the Platonists say, the picture (eidolon in greek), that is, the mirror-image, the reflected form of the Uppermost or the Lowermost in a Human Being.

The destiny of the Soul is essentially dependent upon the personal Spirit of that particular Human Being, with which she is connected. Without that communication she cannot make a connection with the Uppermost. This Spirit of the Human Being is,

66

therefore, the true focal point of the middle Nature, upon which everything is dependant. Therefore, the first and foremost work of the Alchemists is the cleansing and exaltation of the Spirit, which the Hermetics call Mercurius.

The modern Philosophers and Psychologists do not recognize over and above the common conceptual thinking of a Human Being, a higher, transcendental ability of the Spirit. That is, ever since Kant disputed the existence of the pure intuitive intellect, yet Goethe made use of this in his poetic visions.

However in reality, Kant, with his one-sided conceptual intellectual comprehension, unwillingly confirmed, the Truth of the Apostle Paul's words, when he said:

"But the natural man receiveth not the things of the Spirit of God, for they are foolishness unto him: neither can he know them, because they are spiritually discerned."

– (1 Corinth. 2, 14)

The Apostle explicitly assures that the believer did not receive the World-Spirit of the common Human Being, but instead, he received the divine Spirit, which by itself explores the Depth of Divinity.

According to the Words of the Apostle Paul, when a Human Being is not "spiritually appraised," that means as long as a Human Being within himself has not directed his Spirit upon the Higher (the Divine), that is why he will not be able to free his Soul from the bondage of the lower and the demonic. Therefore, it is the first task of the Alchemist to draw his Spirit out of the earthly and the worldly, to "extract" and to "distil above the alembic," that means to lift over and above all ratio into the transcendental and into the Divine.

However, it is not of much use to have only the Spirit, the fleeting white, mercurial Lily-Liquid ascend. It is more important when it comes to the Magnum Opus, that both principles, the

67

Spirit and the Soul, the white mercurial and the red sulfuric Lily-Liquid rise together simultaneously above the alembic. That is why it says: In the kleine Bauern" (little Farmer):

> "Govern the Bath with such Warmth, that both Natures ascend simultaneously, otherwise, if there is only one, you will never enjoy, nor partake in its scent of Balsam."

However, it is not so easy to lure the Red Lion out of his earthly prime-material den, because it is the masterpiece of the whole hermetic art. The Red Lion is locked up in the Inner of the Salt-Crystal, like in a prison, so that he is weak and incapable of freeing himself through his own power. Therefore, a special key is required to open the crystal prison of the bound Soul and to dissolve her earthly chains.

This key is the general-dissolvens of the Alchemists, the Universal Menstruum, the secret Spirit of Vinegar (Spiritus aceti) of the Hermetics, also called AZOTH or Alkahest.

First, we must bring the Light into our microcosmic chaos, which will divide the chaos into two parts, one being active and one being passive (suffering), one volatile and one fixed, an upper and a lower water, that is, in Heaven and Earth of the Philosophers, the twofold matter (rebis) of the Alchemists. Paracelsus said: the Lion must be changed into a white Eagle, through the help of Nature and the Artists work, so that out of one becomes two. This separation of the Materia into two halves, was accomplished by the Ancients, in a very slow and laborious manner, this was called "The Aquilisation" (the preparation of the Eagle). Whereas Paracelsus taught the straight and short way, the extractio sanguinem leonis et glutinum aquilae. Because, through this short way, the whole work can be accomplished considerably faster, than through the long way of distillation, through which you can hardly accomplish the preliminary work in one year. Lullius, Philalethes, Welling and also others taught the short way, which alone is the "VIA REGIA" of the true Adepts.

I. The Work of Dissolution.
The Preparation of the Azoth.

"Whosoever is in possession of this water is the master of this art, however, where this mercurius is missing, there the whole art is lacking." — *(Axiomata, 1736)*

The dissolving water, as aforementioned, is not given to Human Beings by Nature, but instead this secret Spirit of Vinegar must be artfully prepared by Human Beings.

The Adepts keep total silence when it comes to the preparation, and when they talk about it, the explanations are very vague. Joh. de Monte-Snyder says this about this subject:

"Even though, for those who are experienced in this matter, I wrote too clearly, more distinct, than any other Philosophus has ever written or brought into the daylight; there is still a little bit, which I have kept from them, namely, how to make out of an unmelted, brittle, mineral Saturno such highly valuable mercurium, and how to bring this into a red Spiritum. This can only be accomplished in one way and by one means, even though I have already ignited this Light. This was done in the hope, that if God does not allow it, then this will not be understood, nor will it be remembered. This is one thing which I cannot publish through print. — *(Joh. de Monte-Snyder, Tractatus de Medicina Universali)*

In order to give the true seekers and the budding discipulis et filiis of the hermetic sciences a hint, we will quote Monte-Snyder:

"The Mercurius is also pure heat and fire, which keeps his corpus likewise in a constant flow; his fire however, is a mineral, metallic fire, which burns, but it does not burn up, and it can only be ignited by the Philosophis through consummation and amore Veneris etc., and after that he is prepared to warm up the cold Saturnum, because from him he has to expect his coagulation and the Philosophi say: The Coagulation Mercurii is found in the Saturno."

Monte-Snyder does not mean here, the common, but instead, the one prepared according to the art and fundamentliter reversed Venus, which turned into a mineral. Out of this an aqua Mercurialis is made, and with the help of the Sal Ammoniac of the Philosophers (the flying dragon), is being changed into a secret fiery Spirit, into the Menstruum Universale.

The invisible Mercurius or Spirit will initially be drawn out of the darkened Saturno as perspiration. The black earth coagulates the invisible Mercurium, whereas the Mercurius in the water dissolves the black Saturnum. Through frequent ascension and descent of Nature in constant warmth, out of the black Saturno animato becomes a transparent pure and white salt. Out of the black Raven becomes a white Swan, a new illuminated spiritual body.

In fact, everything is dependent upon the preparation of the fiery Mercurial Water or Azoth. It is one of the greatest wonders in Alchemy, to distill the cold watery mundane Spirits into a fiery, lucid and penetrating Spirit, so that he is capable of radically dissolving the dark Salt-Corpus (wherein the Red Tincture is enclosed), and to change the hard stone into a viscous Water, namely into its materia prima.

"Igne Nitrum Roris Invenitur!"

In order to transform the common World-Spirit or Mercury into a fiery, penetrating and dissolving Spiritus, we must first separate him from his external phlegmatic aqueousness, namely, to separate everything that is superfluous and foreign.

Then he must also be separated from his internal impurities. In order to lead him back into his prime Nature or being, into the upper heavenly Light-Water, all connection through which he is bound to the Earth must be severed.

This process was portrayed by the ancient hermetic Philosophers in their writings in different ways, as it was shown allegorically through the distillation of the Spirit of Wine. Just as the sweet juice is pressed from Grapes, which then ferment and become wine, and through distillation becomes a volatile Spiritus or Spirit of Wine.

Furthermore, through the transformation of the sour tartar, it becomes calcareous earth and when distilled with sulfuric acid, it becomes the acrid Spirit of Vinegar.

This is how (allegorically) the philosophical Spirit of Wine and Spirit of Vinegar are prepared. First the juice of vineyard is artfully drawn out of our Subjecto (our philosophical wine) which is the root moisture or the virgin earth, which contains all three principles: Mercurius, Sulphur and Salt.

As soon as we have this virgin earth, this terra adamica, this spermatic Mercurial-Water or Materia-Proxima, then through philosophical distillation or through ascension, we draw out the fleeting Spirit, our white Eagle. This ascending Spirit encompasses within itself the heavenly Light seed, the Sal coeleste or the heavenly Sal nitre, and descends again with the Dew of Heaven, because he is attracted magnetically by the Virgin Earth, which is loosened by the water and swells up.

The Spirit Water enters into this loosened Earth (our Azoth), which contains within itself the Light of Heaven, which is a volatile astral Salt. This astral seed joins now with the Fire of the Anima, which is located in the inner of the virgin Earth and tinges same, that is, it ignites into Light so that the whole Soul is illuminated through the heavenly Light of the Eagle or Spirit, like through a descending lightening bolt. The ignited Anima is driven up again to its heights with the help of her two suns; the empty Spirit Water (or Phlegma) as an Eagle, where in the upper regions of the air or the Heaven it fills itself again with the Milk of the Sunlight, and with this descends again to Earth as Heavens Dew.

Since the Azoth takes, each and every time, a little bit from the substance of the fixed fire up to its heights through its volatility, it becomes more and more acrid and fiery, until it finally reaches such fieriness that it radically dissolves the firm Salt-Corpus.

The whole Heavens Dew will not be coagulated to Light-Spirit-Salt in the fat of the Earth, but only the 30th part of the Azoth or the Spirit of Vinegar. Therefore, we require for the dissolution of our materia, a large quantity of the Spirit of Vinegar, since 10 Eagles or parts of the Azoth only dissolves one part of our virgin Earth.

That is why it is necessary to be imbued so many times with the Spirito aceti and a weekly gentle steaming of the Phlegma, until the subtle spiritual part in the body of the salt is fixed.

There is a foreign impure fire in the Sulphur of the Salt that eats up or corrodes the watery Venus or virgin Earth, and at the same time, it wants to change it into an impure Sulphur.

However, the heavenly Light, which is concealed in the watery Nitro of the Azoth does not allow this. The Heavenly Light that joins the male tincture, the pure fire of the Sulphur and which is effective in its earthly Nitro, separates the coarse burnable Sulphur and cleanses the lower Central-Fire, that means the Anima.

The Water is the Vehiculum of all influences, since all vapours of the earth are led by it into the air, and everything that comes from Heaven descends into the Earth with the Dew-Water. Heavenly and earthly influences, the upper and the lower central sun, meet in this manner in the Water-Vapours.

Vulcanos, the subterranean sun, drives the oily saltwater (root moisture) up in vapours into the upper regions of the Air or the Heaven. Here, it is impregnated by the astral influences of the same, especially by the rays of the sun and the moon.

It then descends back again to earth as Dew-Water, pregnant with the heavenly seed which it softens and dissolves, and in their

Inner, the Central Sun (from the Sulphur) coagulates the Heavenly Light contained in the Azoth, because the Mercurius (Spirit) dissolves the Corpus; the Sulphur (the Soul) however, coagulates the Spirit to a Corpus.

The Eagle (Spirit) is therefore the first part that is led into the heights, Hermes subtile ascension, the Spirit of Heaven, the Fontina Bernhards.

As soon as the Eagle has flown, the body or the green Lion remains behind with the enclosed Soul, the Lion's blood or Gold lustre.

The Sages say: that the whole art lies within that their Gold or Sulphur also rises with it above the alembic, this does not happen with the first distillation of the Mercury.

The blood of the Lion, the sought-after Red Tincture or the Seed of Gold is stuck too deep in the Earth and bound to it, the captured Soul must, through the descent of the Heavenly Eagle to Hell, be liberated from the darkness. The Azoth or the Red Eagle is also the sought-after mediator between the White Eagle and the upper Light-Spirit and the wingless earthly Dragon, that is between the divine Spirit and the earthly body.

This is why, the Adept Monte-Snyder said this:

"The Azoth or the Cross will be understood through the vinegar, & hoc vice versa; through the Cross the Azoth, the Saviour will be comprehended anew, who from up above settled down on this World, and moved above the Water as the Spirit of God of the Salis sive Vitrioli & Nitri, out of which the same is born and extracted. With all this, I could have easily helped you, but I am not allowed by the Powers from up above to disclose anything further in writing."

— (Joh. de Monte-Snyder,
Tractatus de Medicina Universali)

While the Azoth is dissolving the Earth through its fiery water, the Earth coagulates the Azoth at the same time; to be more specific, the most subtile part of the Azoth, so that out of two, a peculiar Middle-Substance comes into being, a thick water or a thick blackish-reddish oil, the materia prima metallorum.

This Mercurius duplicatus is now the blessed Blackness of the Philosophers, the black Raven or Laton of the Alchemists which must be whitened.

In this Ravenshead or Mercury Water, the Sun and the Moon (Soul and Spirit) are being eclipsed (darkened), that means they no longer can be seen.

Because when our Earth has become Water, that is when the dry Fire (Body and Soul) is being dissolved through the moistness (Spirit), then the warmth begins to become active and begins now to act in the moistness. Through this, it bears a blackness in the mixed materia, which are called Sol and Luna. Whenever the Fire or the Anima is drawn out of the body, this always occurs under the sign of Blackness or Darkness.

This then is the Sun and Moon Eclipse (occulation), which means having died or passed away.

Among the Seekers prevails mostly an erroneous concept, in regards to the nature of the Blackness and the alchemistical colours, as well. Therefore, we will endeavour to explain the nature of these in more detail.

Light and Darkness are the primordial principles of all things. All visible things of the three Kingdoms of Nature are created out of a fiery Darkness and a watery Light. The Prophet said:

"Since eternity, Light and Darkness were in God, and both those things came out of God." – (Isaiah 45-7)

Before Creation, the Darkness was not revealed, because of the magnificence and sovereignty of the Light; it was instead concealed in the Light of Wisdom and the Holy Trinity. The Light had to first give birth to the Darkness, so that the eternal Wisdom could view itself therein, just like in a mirror, otherwise God could have only revealed Himself to Himself. The Light is the first

Materia of all things, and all Corpora inclusive of the Body of Human Beings will only then be fully restored to their first dignity, when the Light will again shine out of them from within.

> "Trismegistus, in the countenance of Creation, first saw a lovely and joyful but intermingled Light. Following that, a dreadful, sorrowful Darkness appeared and moved downward; it descended from the Light like a cloud from the Sun. This Darkness was a certain thickened Water. After that, the Holy Word came forth out of the Light, moved upon the Water, and made all the things out of it."
>
> – *(Thomas Vaughan, Aula Lucis)*

In Creation, the Darkness is separated from the Light, as Eve is from Adam. As the woman in the separation has retained much of the male Essence, that is why the Darkness in the separation has retained much of the Light-Essence, because they cannot be totally separated from each other.

That is why the Darkness yearns for the Light as for its true life, because originally they were one and dwelled together in the greatest of harmony. Also, the Light yearns for the Darkness, so that it may reveal and perceive itself therein.

When now, these two extreme principles are being brought in this manner into harmony with each other in a middle, third principle, so that the materia of these becomes transparent and illuminated, that then is such a Being, that can temper all ailments and disagreeableness, may they be human bodies or all the other bodies, which are created out of Light and Darkness.

It should be understood, that when we are speaking about these incomparable Beings, we are talking about the Lapis Philosophorum.

Originally, the Darkness was not evil. It was good because the Light was in control of the Darkness, and it was therefore, only a gentle impetus and reason for the effect of the Light, so that God's works could be revealed through it.

75

It seems, that the darkness became evil only in the lower creation, because those creatures who were created in a total equilibrium of Light and Darkness, aroused and multiplied the Principle of Darkness within themselves, and in this manner they let the Darkness gain dominion over their Light. Through this, it began to suffer.

That is God's Lamb, or from the Beginning of the World, the crucified and strangulated Light! — *(Revelation 13-8)*

"Divine Wisdom is a Light and this Light is the Life of all Creatures, and Life is a Tincture, and this cannot be encountered fixed and constant in any, but the metallic Nature. That is why the Sages followed this Light out of the occult Wisdom step by step, and they found it, in an unsightly dark body, where nobody would have easily looked for it. They then liberated it with the upper Light, as its origin, tinges and multiplies, and then it can be used as a remedy to strengthen their natural Life."

— (Microcosmic Vorspiel
= Excerpt from the Microcosmic Prelude)

Adam (the Human Being), was originally the Quintessence of the Whole of Creation, the small World (Microcosmos), the image of God and the large World. Before the fall, when Adam carried God's picture in an obvious manner, he was in Paradise. After the fall, the Light in him became darkened through the predominance of the earthly element. What once was Light, became a shadow, that is, it is now a coarse dark body. If the Human Being wishes to again reach Paradise, then he must dissolve the dark shadow and draw again the hidden or occult Light out of the Darkness. That is why the Alchemist said:

"Son, draw the ray out of his shadow!" *— (Hermes, Cap. II.)*

If we, therefore, through the subtile divine Light will be born again from up above and will be tinged and illuminated in our

Soul, we then can cleanse the lower Light from its curse which is hidden and darkened under the darkness and in this manner bring forth a new microcosmic birth.

Originally, the Light was not darkened by the Materia, because in Paradise the Body of Adam was a pure, transfigured, transparent, crystalline body, like a living diamond. It was so pure that the Light which lives within him could radiate to the outside like a Sun.

Adam himself was the Light, the Son of the Sun, and the fluid diamond body which surrounded him was the with him united moist lunar Eve or Venus.

However, through the ignition of the luciferous Fire, the luciferous Fire gained the upper hand in Adam, then began the arousal of the worldly heat in him to dry out and coagulate the moist nature of Venus. So that the Earth came forth out of the crystalline Water, through which the Light was wrapped in Darkness.

Adam's immortal Light-Body was now concealed in this terra damnata, that is in the coarse earthly body. Yes, he himself became powerless and dark, because his noblest powers were swallowed up by the earthly Body. The "Tree of Life" became the Tree of Death.

At the same time, while the Body was materializing, a division of the sexes occurred. Originally, in Paradise the Human Being was androgynous (Man and Woman), Adam-Eve in one form. After the fall into the coarse Materia of the lower World, in accordance to the outer body, the Human Being was divided into two sexes.

In those Human Beings where the solar Light-Principle was predominant, the outer body took the shape of a male. On the other hand, in those Human Beings where the moist lunar Principle was predominant, the earthly body took the form of a female.

The later urge of the sexes towards each other is, in reality, a dark and misguided aspiration towards our own rebirth and renewal of our Being. It is, in reality, an aspiration towards a unification of Sol and Luna in the Human Being, that is, to reclaim the once blissful State of the immortal paradisiacal Hermaphrodite.

Basically, the Human Being wants to procreate itself anew, in order to live eternally like the Angels. Yet we produce only mortal descendants because the two Principles, Sol and Luna (Fire and Water), became separated in the Human Being through the fall. Adam (the Soul) no longer had access to the "Tree of Life," that is, to the Powers of Eternity of the crystalline, mercurial life-waters, the magnificent white tincture, out of which an imperishable Light-Body was once built, since he turned towards the deadly fruit of the Tree of Knowledge and became desirous of the mysteries of the lower, materialistic World.   – *(Genesis: 3, 22-24)*

At this point his eyes were opened, and he became cognizant that he was naked; in the coarse, earthly sense, he was without clothes. That is how he lost the immortal, full Light Body of Paradise, and received the grey, mortal "Coat of Skins."

– (Genesis: 3-21)

Since, in our Subjecto, the coarse Darkness by far surpasses the Weight of that in his Inner concentrated Light, it is therefore, very important when it comes to the hermetic Work, to separate, by means of the Azoth, the superfluity or excess of the Darkness from the Light artfully and subtile and to increase and improve the enclosed little lower Light through the greater upper Light, qualitative and quantitative, to exalt it again to its first dignity and perfection.

Originally, Adam was created out of the Quintessence of all the Elements, that is out of the materia prima of all things. The Sages call this the Mercurius Philosophorum, that is a dry water.

In the beginning, through Divine Wisdom, the Great Whole World was created out of this materia prima (Chaos). This materia prima is with the Human Being one nature, but with this difference, that this Primordial Matter expands in all creatures, but in a Human Being it concentrates. That is the true Self-Knowledge, the highest Study in hermetic Science.

Therefore, you must be mindful in the new microcosmic Creation, that is, in the Magnum Opus of the Sages, that all Elements of our Subjecti, as well as the lower and the upper (Light, Air, Water Earth, Fire, fleeting and fixed Spiritum) be made into a philosophical Water, or respectively into Wine, in which Light and Darkness, Death and Life are together, as it was in the primordial Chaos of Creation.

When that is accomplished, this Water must be gently separated from the Earth, in the divine Fire, so that the subtle Earth which is hidden, philosophical Wine or mercurial Water comes forth.

When all the Water (Spirit) has been separated from the Earth, and the Earth is covered with Darkness, then the heavenly Seed will be sown therein. This Seed will later prove its blossoms through a multitude of colours, and will enter into a white and eventually into a red immortal fruit; this is the highest goal of all of Alchemy.

This Darkness is the desired and blessed philosophical Blackness which has concealed within itself a delicate Light. It is a virgin, a totally uninhibited Earth that has never before bore anything, but from her heart desires the Light. This new Earth upon which the whole Heaven has his eyes, is the true Gold-Magnet who draws towards itself the Seed of Light of the heavenly Sun and the Moon, and within itself solidifies into a Corpus. This Earth is only a subtle shell of the Light, and a matrix wherein the upper heavenly Light clothes itself and becomes visible.

But, in this Phase of Dissolution, is that in the Chaos by the Matrix conceived heavenly Light is not visible yet, instead it is still veiled by the black Darkness of Death.

79

This condition of the philosophical Blackness is described by the Alchemist John Pordage (some consider him to be the English Jacob Boehme) with the following words:

"Because, through this, the delicate tincture, this delicate Child of Life must descend into the forms and attributes of Nature, so that it may suffer and endure the temptation, so that it may continue to exist; by necessity, it must descend into the divine Darkness, into the dark Saturnum, wherein no Light of Life can be seen. It must be held captive therein and be bound by the Chains of Darkness and it must live from the food, which the prickly Mercurium will give it to eat; and that is for the divine Life-Tincture nothing else but dust and ashes, poison and gall and fire and sulphur. It must enter into the fierce, angry Martem from which (as in the case of Jonah in the Belly of Hell) it is being devoured, and to feel the curse of the wrath of God; and also being tempted by Lucifer and Millions of Devils, who dwell in the Wrath-Fire Characteristic.

Through this, the divine Artista sees the first colour in this philosophical Work, wherein the tincture henceforth appears in its Blackness; it is the blackest Blackness. The learned Philosophers call this their black crow or their black raven, but also the blessed and blissful Blackness. In the Darkness of this Blackness, the Light of Lights is concealed in Saturni Characteristic or Quality, and in this Poison and Gall concealed in Mercury is the most exquisite Remedy against poison – the Life of Life. The blessed tincture is concealed in the wrath or fury and the curse Martis.

At this point, it seems to the Artist that all his work is for nought. What happened to the tincture? Nothing is being revealed, nothing which can be seen can be recognized or tasted, other than Darkness, the painful death, a hellish, frightful fire, nothing but wrath and the curse of

God. It cannot be seen, that in this putrefaction or dissolution and destruction of the Tincture of Life, that in this Darkness there is Light, in this Death there is Life, in this Wrath and Fury there is Love and in this Poison there is the highest and the most exquisite tincture and remedy against all Poisons and Ailments.

The ancient Philosophi called this work or labour, their descent, their cineration, their pulverization, their death, their putrefaction of the materia of the stone, their corruption, their caput mortuum.

Do not despise this Blackness or black colour, but endure therein with patience and suffering and in silence until the Forty Days of Temptation have passed, until the days of Suffering are completed. Then the Seed of Life will arouse itself to Life, come to life again, sublimate or glorify, change itself into white, cleanse and sanctify itself, and give itself the redness, that is to transfigure itself and achieve permanence."   – (John Pordage to Jane Leade)

The german Alchemist Siebmacher writes:

"This Stone's Genus is everywhere.
His Conception occurs in Hell.
His Birth he has on Earth.
His Life he leads in Heaven."

The alchemistical work actually takes place in three Worlds; the lower, darker world, followed by the middle, paradisiacal world, and eventually in the upper, heavenly world.

Each and every one of the three Main-Works of the Maganum Opus takes place in the appropriate sphere.

The first work of the so-called preliminary work ends after the dissolution, through which the old world is being drowned by the flood; therefore, in reality in hell, that is, in the Darkness of the Underworld.

81

The christian Mystic Angelus Silesius had knowledge of this, when he said:

> "Christ, for once you must be in the Abyss of Hell. If you do not go when you are alive, then you must enter when you are dead." — (*Cherubinischer Wandersmann*)

Jacob Boehme said this:

> "Forthwith after several severe Storms, my Spirit broke through the Gates of Hell into the Innermost of the Deity, and was there embraced by Love, like a Groom embraces his Bride." — (*Aurora*)

With the attainment of the Ravenshead of the Alchemists, the first work of the Magnum Opus is concluded, because in the blessed Blackness is the beginning of the Birth of the paradisiacal Son of the Sun already present.

II. The Main Task or the Main Work.
The Task of Cleansing.

"This firstborn Son of Nature, even though in his centro his substance is pure, must be renewed and be born again through Water and Fire; through the separation of the Earthly from the Fiery, the Coarse from the Subtile, with one Word: The Impure from the Pure. The cold moisture, which is mixed with earthly heavy things, must be withdrawn in order for the dry warmth to enter. That is how it ascends from the Earth towards Heaven, from Imperfection to Perfection. Nothing can attain heavenly Perfection unless firstly, the imperfect coarse mortal crust has been shed, which is of cold characteristics and the cause of death, whereas warmth produces Life. It is therefore, Nature's Rule that her subiectum has to endure and go through a dark Blackness of Death, and through this must expect the clear white Immortality, the renewal of Life."

– *(Nuysement, vom allgemeinen Weltgeist)*

The Killing of the Corpus of our Subjecti should not be understood in the manner that the same is being destroyed through a coarse Fire and in this way it becomes lost. Instead it undergoes in the manner of a seed of subtile putrefaction, through which is being taken the old specific and temporal form, while at the same time out of him grows a new universal and immortal form. This new form is considerably more noble than the previous one; and so different from the first fragile form, as the highest rectified alcohol seasoned by its Salt, is different from the watery Substance of Wine, out of which it was extracted.

Only those who know the Salt of the Sages and its preparation will understand all of this. In the beginning, this Salt does not have the Salt-Form, instead it is black, impure and formless like coagulated blood, which must be cleansed and whitened.

The Salt is the first corpus, through which the materia prima becomes comprehensible and visible. It is a pure virgin earth, in which the World-Spirit inverts itself through the coagulation of the water, because nothing is being coagulated, other than the Salt alone, and nothing but the Salt is being dissolved. Everything that is made out of the four Elements, can be changed into a Salt again.

Whenever a body putrefies, then ash and dust remains; within this, an exquisite Salt is concealed. Originally, all bodies were made out of this Salt.

The beginning of things was a Water or a moist Nature, upon which the Spirit of God moved. Then a subtile Earth took form in the Inner of this Water.

The Alchemists differentiate between two kinds of Water or Mercury, namely: one is airy and ascends, and the other that coagulates to Salt and basically becomes fixed.

Paracelsus signifies with Salt "The Centre of Water" wherein the Metals should die. When this "sal circulatum" is being distilled many times, it looses its solidity and becomes vitriolic Water (materia prima).

As long as this Salt of Nature is spread out, but stuck in its limbo or chaos and at the same time, it is still stifled in its water, it then does not appear in any form and characteristic, other than that of a bitter, salty water. It remains in this condition until, with the help of an Alchemist, through warmth in the separation, the excess water is evaporated and the salty earth appears like an island in the ocean.

This black Earth, which emerges out of the Water of the Deluge, is the Ravenshead of the Alchemists; the Toad which eats up the Eagle or Spirit; our philosophical Saturn which swallows the Moon and retains it in its belly; the Earth of the Sages which eagerly swallows the Golden Rain. In brief, it is our Laton, which is washed and must be baptized seven times in the Jordan.

Out of this black Earth, the Sage should bring forth magnificent Lilies and Roses, and trees, who carry imperishable Moon and Sun fruits. — *(Deuteronomy 33-14)*

In other words, the Alchemist should change the unfruitful, waterless desert (which is located in a very secret place), into a flowering Paradise!

Due to the continuation of the warmth, the parched Earth drinks again all its eliminated moistness, more precisely one white Eagle after the other, and finally becomes dry.

Since in this work of cleansing the Corpus should be drenched, washed and united with the most fleeting Spiritus volatilis, as it was previously done with the Spirit of Vinegar which dissolved the body and made it receptive to such a degree, that it was capable to accommodate within itself, the subtile, heavenly Light-Spirit, which would have been impossible without the procurement of the Azoth.

It should not be forgotten here, that the Holy Ghost was not poured out until after the Exaltation through the Cross. That means, through the Azoth, after the Blood and Water flowed! Just as the first labour of the Magnum Opus was a work of dissolution and Death, the second labour is a restoration of Life and a work of rebirth.

The Preparation of the White Lapis.

When the dissolution has properly occurred, a subtile, black Ash forms on the top of the Mercurio, like a fog above the Water. In Genesis (1-2), it says:

"... and the darkness was upon the face of the deep, and the Spirit of God moved upon the face of the water."

This blackness is the Tincture which we seek! This is our Ravenshead, the black Sun of the Alchemists, which we must cook in a gentle fire until it sinks in the water and remains on the bottom of the container as a thick, black water. This is called Oleum Philosophorum. This is, in reality, a fatty, black powder, an incomprehensible dust, which can only be seen in the Rays of the Sun. This is our black Elixir, which must be whitened.

In this manner, the water received something other than its own nature, so that they can no longer be separated from each other.

There are three operations in every Work:

1. To dissolve the Body with the Spirit.
2. To cut off the Head of the Raven.
3. To make the Black white, and the White red.

Before we take a closer look at the details of the second operation, we should also mention what Paracelsus had to say about these procedures.

He writes in his Handbook the following regarding the Philosophers Stone:

"When your Electrum is broken, then take the Electrum which was made broken and fleeting and set it into a philosophical egg and seal it well, so that it cannot evaporate. Let it remain in an Athanor long enough, until it begins to dissolve by itself without any addition and until an island can be seen in the middle of this ocean, which reduces daily and eventually turns into a cobbler's blackness. This Blackness is the bird, which flies by night without wings, which also has inverted the first Heaven's dew through constant cooking and ascension and descension into a Blackness of the Ravenshead, and afterwards becomes a Peacock's tail, and then acquires Swan feathers, and eventually accepts the most Redness of the whole World, which is a sign of his fiery nature. This production occurs, in accordance with the opinion of the Philosophers in a utensil, in an oven, in a fire, without cessation of the vaporific fire. Thereupon, such medicine is heavenly and complete. Therefore, be serious about this; nobody can comprehend or understand this Arcanum divinum without divine Will. That is why, its virtue is infinite, so that God can be recognized therein. It is forbidden to write more about this secret and it is reserved for the divine Might. Because, this Art is actually a Gift from God, and that is why not everybody can understand it. Therefore, God gives it to whom He wishes, and He will not let anybody extort it from Him with force, but instead, God alone wants to have the Honour; His name be praised eternally. Amen!"

– *(Paracelsus, Manuale de Lapide Philosophorum)*

Paracelsus calls the Raven of the Alchemists, a Bird, which flies at night without wings! This is the up-to-now unaware, dream-like Depth-soul, the transcendental Subject of a Human Being, which now no longer is separated, but instead, it is being exalted to the consciousness of the Spirit and entered into a gentle communication (coagulation) with the Spirit.

This unification with the Divine Spirit is, however, for the animalistic Principle, that is for the lower Astral Soul, the old Adam within us, to a certain degree deadly, he is suffering through a

87

condition of death, where he looses his up-to-now demonic Astral-form and enters again into the antenatal Limbus, that is, into his first creative indifference, into his materia prima! Now he became Earth again, from which he was taken, that is, APHAR min ha-ADAMAH, a red tinctorial Gold-Dust, united with the Dew of Heaven. — *(Genesis 2-7)*

At this point, his holy Earth is still dark, desolate and empty, but the Spirit of God is already moving about it to impregnate the divine Light-Seed which is concealed within it, with the upper Spirit-Water. Previously, during the ascension and return of the Spirit of Vinegar, the materia which was below remained constantly moist and was boiling in a continuing movement and evaporation. Artephius says:

> "The whole Mastery is depending upon the boiling and driving away of the spiritual smoke or vapour, which takes approximately 6 weeks."

First the steamed-out water ascends white, like a Phlegma. As soon as the inner Central-Fire is aroused, the ascending steam turns yellowish and gilds the Glass inside like with subtile gold flakes. These fine gold flakes gradually sink as a yellow powder to the bottom, which little by little turn ash grey, become dryer and dryer, and eventually turn into an incomprehensible pitch-black dust.

From this moment in time, the steaming-out ceases for approximately 6 weeks. Because when the water which escapes through the evaporation is no longer colorless and phlegmatic (but instead coloured yellowish and is sour like vinegar), that is a sign that the salt is sufficiently satisfied and dissolved, so that now, it accommodates in itself its subtile Spirit volatilem.

The fine and fleeting parts of the body have become mixed with the subtile parts of the Spirit, with which they ascended. The coarser parts of the body on the other hand joined the fixed parts of the Spirit on the bottom.

The adamic Earth is now a middle substance between the body and the Spirit, an intermediate thing between the fixed and the volatile. It is not as fixed, that it would be like gold, but also not as fleeting as mercury. Nevertheless, this new body is fixed enough, that it can endure the fire, which is necessary for this work and all the torture and pain of this purgatory of the Alchemists, wherein it remains for 40 days and nights without smoke or steam, as Ripley explains this in detail.

The subtle Spirit (or Spiritus volatilis) remains in the upper part of the glass during this process, since he cannot endure the fire, and when the steam ceases, he will eventually be drawn down through a magnetic force.

The whole secret of this process therefore consists in the proportion of the glass in comparison to the materia, because it has to be large enough, that its inner cavern can accommodate the appropriate amount of water which comes from up above. That is why, the glass should not be filled with more than one-third of the materia.

As soon as the Blackness appears, the boiling must be stopped. This discontinuation of the continued digestion or calcination is very important at this point of the Magnum Opus. These ascensions and descents or circulations of the Spirit were only the means and not totally without danger of bringing about the condition of their external darkness, the void and desolate desert, that is, to bring about the materia prima.

An exaggeration of the spiritual exercitation would only disturb the inner equilibrium of our subjecti and it would be subjected to unnecessary disturbances. It is very important that during this condition of blackness or the Ravenshead, to remain with total calmness of Heart in this Abyss of the Hermetics, until the "40-Day" examination and purgatory have been passed through.

According to the old Egyptian Mysteries, the Adept must pass through 12 Kingdoms of the dark Underworld, whereby he has to change garments twelve times, so that eventually by morning, adorned with the white raiment of the God Osiris, he can be shown to the crowd, as the "Son of the Sun." In the Persian Mysteries, the lowest degree, was that of the "Raven," the fourth degree, that of the "Lion," and the seventh and the highest, that of the "Eagle"!

With the appearance of the Blackness, the first rotation of the Wheel of Nature is concluded (that means, the first rotation of the Magnum Opus). This is followed by a second circulation, through which our materia is being changed into a white Silver-lime.

But before, we proceed to the next operation, namely to cut off the Raven's head. The Sages teach: "Deprive the Raven – his head." That means, you should separate the moist Mercurium from the Blackness. Otherwise, the dryness cannot overcome the moistness, and then the Blackness would become fleeting with the moist Mercurio.

In the beginning, when we compounded Sol and Luna in the moist Mercurio, we required a great quantity of moistness, otherwise, it would not have been possible for the moistness to overwhelm and dissolve this dryness.

Now, we must place the weight in such an order, that there is now more dryness than moistness, because now, our materia must be cleansed and coagulated. Therefore, the dryness must now be given all the aid, so that it becomes dominant over the moistness.

In the steeping which follows with the Spiritu volatili of the Blackness, you should not all at once, add too much moistness, otherwise, the seed will become too moist and the dryness would not be effective, and a coagulation cannot occur.

The Sages' advice: "Make the red Laton white, and after that burn the books, so that your hearts will not be broken."

Now, there must occur a new steeping with the aqua secunda or the Spiritu volatili, that means, with the white Argentum vivum, as it was done before with the Spirit of Vinegar.

When the glass with the Blackness has stood for 8 days and nights in a gentle fire, it is then removed from the fire. Then the Blackness is being moistened with the Virgin's Milk, drop by drop, and in such a manner, that the dryness continues to keep the upper hand.

When the Sulphur is being steeped with too much water, it then again turns into a yellow oil (Aurum potabile). It resolves and floats on the Mercurio, and for this reason a union cannot occur. The Sun should not float on the Moon! We should follow the example of the potter, who slightly moistens the clay, just enough so it becomes cohesive.

Then the Blackness, which has been moistened in this manner, is left standing for 2 days and nights without any external fire, and the materia is left with its own fire to work with, until the quarrel between the two Central-Fires has been settled and the materia settles again on the bottom. Following that, the glass with the moistened Blackness is again placed into the oven. However, it must be placed into such a gentle fire, that the Mercurius does not ascend and fly away as an Eagle. The materia remains in this manner for another 8 days and nights, until it is totally dry and the Spiritus becomes coagulated within.

These little steeping processes are being repeated every 8 days and after the steeping process, it will remain for 2 days and nights without fire in the ashes, until the whole Mercurius is imbibed by the blackness and the dry Earth has drunk the whole Spirit, which is approximately 50 times as much as the Corpus.

At the same time, the Corpus increases through the steeping process and each and every time this is done, the Spirit dissolves with the blackness and with it becomes an ash, which however, is made more subtile and whitened through the Spirit.

The larger the Corpus becomes, the more Spirit can be added each and every time, as it can endure the heat longer. That is how the Gold-Embryo continues to grow, through the nourishment with the Virgin's Milk. Nature insists that a certain time be taken to accomplish this work, therefore, the steeping and coagulation process should not be hurried.

Therefore, the Spirit should not be urged too much, because it is a fleeting substance, which would otherwise break the receptacle and would flee.

> "Because, when your heat surpasses the natural measure, then you have aroused the wrath of the moist Natures, and they will stand up against the Central-Fire, and this will stand against those, and in the chaos a dreadful disruption will occur. The sweet Spirit of Peace (the true eternal fifth Being) will leave the Elements and will leave you and them behind in confusion, and they will not join this materia again, as long as they are in your cruel destructive hands. Be aware! Do not become Satan's apprentice, whose only aspiration is to spoil and destroy. I speak the truth from experience." — (Eugen. Philaletha, Magica Adamica)

After the mercury dissolved the Sulphur into its materia prima, then made the green Stone first into a yellow Stone, then made it into a deep-blue viscous Water, and eventually coagulated it into a moist, oily, pitch-black lime, Saturn began his regiment with this blackness, that in its materia had the appearance of a black scaly Toad, which lasted during the whole time of putrefaction. As soon as the materia, through its own inner fire, dries up totally to an incomprehensible dust, the ice-grey colour of Jupiter appears as a sign that the sovereignty of the water is over and the kingdom of the air begins to be effective.

Now the materia takes on the form of ice, like a crystal. This jovial Regulus is now our Stone Onyx studded with Rubies (the seven red Rubies of the Stone signify the 7 steepings of the fiery or golden Water) (see "Sun from the East" 1783, page 65). That means the new Gabritius, which still has to be washed 7 times or be baptized in the philosophical Jordan.

Following the Sovereignty of Jupiter, Venus now becomes the

green of the new Kingdom of the Air, which now nourishes and washes the developing material Sun Child with its Virgin Milk.

This is followed by 7 washes, rinses or steepings, to which the Spiritus volatilis is added 7 times to the materia. When this occurs, all the colours of the rainbow appear, the so-called Peacock's tail, as a sign that Venus has begun its rule.

The added Spiritus volatilis dissolves the Laton, each and every time, into a milk-white fluid, which is also called Lac virginis, This is dried up again and coagulated into a fixed salt. Pay attention when this occurs. Whenever you rinse or wash, the Fixum must always surpass the Volatile by weight, so that, at all times, the materia outweighs the Spiritu volalitii, otherwise the coagulations would take too much time or would not take place at all.

Great attention should be paid to the fact that no rinsing or washing take place, nor should there be new Spiritui volalitis added, until the materia is dried up well and stabilized, or that the volatile salt, namely the Mercurius volatilis, through the fixed salt is likewise changed into a fixum.

The colourspiel of the Peacock's tail is as follows: firstly, many colours appear, all in disarray. For example, white, red, black, green, yellow, grey, blue, everything is mixed. Then, all these colours disappear again, and one main colour after the other appears in the following sequence: black, black-blue, dark red, highly red (this colour however is not yet a perfect colour), orange, lemon-yellow, ash-yellow, green, then blue-white, and finally after cooking it for a long time, the highly white colour appears. Before the whiteness, the Laton is green!

When this wonderful green colour appears above the renewed adamic earth, which is now called "The Sage's Vineyard," then this blessed green is a sign that the materia, through the Will of God and His Omnipotence, has reached a new growing life!

Noah planted his vineyard soon after the flood, which blossomed splendidly and brought forth grapes. After this event, there is no longer the need to fear that the work ends up in failure.

When this philosophical vinestock begins to blossom and brings forth delicate green grapes, then the time is close to an abundant vintage.

That is why, at this point, being full of hope, the Philosopher's call out: "O benedicta viriditas, gyrans per universum, cujus centrum ubique peripheria vero diffusa per omnes Naturae Abyssos."

The colorful rainbow, that appears above the new Earth, is a sign that the flood has reached its end, and there is peace between Water and Fire. It is the mystical Jacob's Ladder, which stands between Heaven and Earth, and on which God's Angels ascend and descend.

All colours ascend and descend in the glass approximately 7 times. Every colour is the symbol of a certain new energy and virtue of the developing Sun-Child. The colour changes originate with the inner sulphur, as the Originator and the Producer of all colours. They come into being out of the fixed half, however, still to a certain degree volatile Spirits, which are on their way to becoming fixed.

Yes these colours are, to be more explicit, the expression of a very specific condition of the inner Central-Fire. The Blackness, as the saturnine Colour of Darkness and the Underworld, followed by the ice-grey Colour of Jupiter, as a symbol of the mystical twilight of a new day of creation, whose Light slowly begins to ascend in the East.

With the multi-coloured and the magnificent green colour of Venus, the inner sulphur attains new energy to grow and become more and more alive. Eventually it wins, after the green colour changes into a wonderful Azure-Blue, the dazzling white colour of the full Moon, the Light-Raiment of the resurrected OSIRIS.

This dazzling white colour signifies the Union of the Male

(Sulphur) with the Female (Mercury), so that they form one Body, the Hermaphrodite.

The sought after white Nature-Sulphur, the flying Dragon of the Alchemists, the DIANA of the Sages, the Queen of the Philosophers, the born-again Moon with his Doves, the new paradisiacal Light-Body with his from-above radiating Spirit-Forces.

Since the Eagle is the symbol of the ascending Spirit (Isaiah 40-31), so is the Dove a symbol of that descending divine Spirit from up above. That from now on, like an illuminated Aura or a fiery tongue, remains suspended above the head of our Subjecti, which was baptized 7 times in the mystical Jordan.

"This is a noble step from Hell to Heaven, from the bottom of the glass to the top of splendour and power, from the darkness in the blackness to the dazzling whiteness, from the highest of poisons to the highest of remedies."

– *(Ripley)*

The same as the Souls soon after their transitory pains reach paradise, where there is eternal Life of Joy, that is how our Stone, after its darkness is being purified in the purgatory of putrefaction, and the elements are being joined together to a Quintessence, that means, to the white Elixir of great Virtue, without antagonism.

The body has become so subtile, through the steepings with Spiritu volatili, that he ascends with the Spirit at the same time, and rises with him into the paradisiacal Kingdom of the Air. The Materia has become an incomprehensible white powder, a white flaky earth (terra foliata alba). This is the first Whitening of the Laton.

If, at this point, you wanted to ferment fine silver, one part of this Elixir would tinge ten parts Mercury into Lunam puram. It would, however, be a great pity if this work would be destroyed before its fruit is ripe. That is why the Sages advise not to act like

95

the ignorant Alchemists, who abandon their work at this point. This is actually the beginning, and they proceed hurriedly to the projection. Then when they accomplish little or nothing, without hesitation they place the blame upon the Hermetic Art, or even upon the old Master Philosophers, without considering that they were solely deceived by their own laziness and their thoughtless nature.

In this condition of the first Whitening, the Laton is but a Corpus, still totally volatile and mercuric. Since the Stone of the First Order is totally purified and has been brought into a spiritual, mercuric substance, it does not yet possess the energy to tinge.

This whitened, spiritualized Corpus has not yet the energy of the tingeing Soul, from which it became separated through washing or cleansing, as it did from the blackness. The blackness or darkness is contained in nothing else other than in the red Sulphur. That means, in the inner Central-Fire, as soon as this is through the solution, which separated our compositum into three principles (Mercurius, Sulphur, Sal) is being drawn out of his corpus.

First we draw the Spirit out of the Earth, either through distillation or through the short way of extraction, so that out of one thing, two things come into being: The Rebis of the Sages.

Following that, the Spirit dissolves the Corpus and extracts its Soul. This occurs under the sign of blackness and this procedure of the Soul is the same as actually dying.

With the dissolution of the corporeality (in this instance, the Astral-Form), the Soul has not completely left the body, but still remains to be connected with the dissolved subtile substance of that same body (Ashes, Mumia, materia prima), which therefore also appear black.

In the subsequent work of the Whitening or cleansing, nothing else occurs, but the step by step separation of the subtile Ash-Salt; this academic earth; this dust APHAR from the black

Sulphur ADAMAH. As soon as this magic Earth APHAR, which forms the seed of the new Spirit-Body is separated from the lower Central-Fire or the black-red Sulphur, it becomes white like snow, because, according to its own Nature, this holy paradisiacal Earth is as bright as the heavens and as transparent as the purest crystal.

Through the first Whitening of the Laton, the holy paradisiacal Earth has again attained its original paradisiacal condition.

The pure heavenly Spirit joined with the pure substance of the new body and, in this manner, formed a new transfigured Spirit-Body. The Sages call this their DIANA, their QUEEN or white, natural Sulphur, that is the white Stone of the first Order.

However, the Alchemists consider this white Lapis only to be half a birth because it is totally lunar. The fiery solar principle, the Central-Fire of the Anima from which it was separated, is at this point, still totally missing. And, since this anima is not yet cleansed and exalted, that means it is not tinged, the work can still be destroyed even after the first whitening, when the Alchemist endeavors too eagerly to commence with the projection (with transmutation of the metals), and when he leaves the work at this point, where it actually should begin.

Whosoever has reached the point of the first Whitening of the Laton, has reached the spiritual re-birth, but it still lacks the psychic or astral and physical re-birth!

The Preparation of the Red Lapis.
"Sow the Gold into the White Foliated Earth!"

In the ensuing operation, we are chiefly dealing with the re-unification of Sol and Luna, that means to unite again with each other Soul and Body. The Tincture or Soul cannot be newly born in its old impure body, therefore, the Soul must leave the dark Abyss of the damned, dead, insoluble Earth and she (Tincture or Soul) has to take on a new transfigured body before she can unite herself with the heavenly Light.

That is the reason why it was a must, that through the dissolution out of the dark terra damnata, the Salt of Nature which was hidden and enclosed therein had to be drawn out to again become a Water, a Limbus, a Quintessence of the Macrocosm, which through the influence of the Spirit coagulated into a new Salt of Magnificence. Out of Water and Spirit everything must be born again. When the Soul is separated from the Body in this manner, the Soul or tincture must be mixed with the Spiritu volatili and circulate with it for approximately 20 days and eventually be distilled out of the Ash-Cupella, so that all impurities which are still present can be cast aside.

Thereafter, the red Oil or the Philosopher's Gold (the Soul) is returned to the new white body drop by drop, so that the magic Gold can nourish itself therein, and thereby receive new energy for growth. That is how the Philosopher's Sun and Moon are united.

"Call our Son of the brilliant Sun, our Virgin Earth out of damascene Fields, our double Mercury-Gold, which no longer can escape the Fire.

This is our secret operation, our adamic Earth, a thing of perfection; that is the end of the work of the mighty Hercules. This is the exquisite Blood of our serpent, the true unification of the four Elements. This is the Secret which the Mighty seek; in our books, you will never find this being revealed."

– *(Fridericus Gualdus, Philosophia Hermetica)*

After the white-foliated Earth of the new Spirit-Body has taken in her Soul as a pure, ruby-coloured, incombustible Gold-Oil, she herself is then turned into a red Lapis. Even though the Stone is, after the first whitening and also after the first red-colouring, still very volatile, and is like a subtle dust, because in this condition, the stone is still more spiritual and mercurial. That is why the white or the red Stone of the first Order does not have a higher strength or potency to tinge, than 1:10.

In this condition, Body and Soul are only loosely connected to each other and with each other; they are not yet permanently constant and fixed.

In order to make out of this a united fixed born-again being, before the conjunction we pour over the re-inspired body (a body which received a soul again) three parts of the Spiritui volatilis, so that the greater part of the Spirit surpasses the smaller half-fleeting part of the body and ascends with it to the heights. Through this, the new Spirit-Body is mixed with the Soul and the Spirit and raised seven times into the Air, through which the Spirit-Body is truly exalted and is impregnated with the heavenly influences until the materia sublimates itself totally clearly, like a star. This is the true exaltation of our Stone, our in the Air born-child, which as soon as it returns again into its seed containing water, coagulates with the Soul and the Spirit into a red-gold sap, which is our Elixir.

III. The After-Work.
The Conjunction or the Unification.

The third operation acts with a quiet insensitive movement, which is called "digesto." That is why the Sages say that this operation occurs in a heavenly oven. The nutriment is now being cooked and changed into the substance of the body, that means, the mercurial Nature is being changed into a sulfuric Nature, and this must take place in a very gentle warmth.

This operation is also called the burial (inhumatio), because the Spirit is being lowered into the Earth like the dead. Also, this work progresses very slowly, and therefore requires a longer period of time.

The reduction is also the composition of our Apollinis or red Mercury and the Dianae, that means the foliated Earth. The foliated earth is also called the fermentation, because within it our fixed, natural Sulphur is liberated through our red Mercury or Azoth and ferments through the degrees of the philosophical boiling to become the permanent Philosopher's Stone.

One must however, also draw out of the materia of our Stone, the red man (or Sulphur), and his white wife (or Sal), which are the two Main-Principles of our Subjecti. That is why the Philosophers say: The Azoth and Ignis are sufficient for this work.

So that the two can enter into one another, into an active life and may unite in the centre, the astral Spirit, which is the means of their unification must step between the two and help to accomplish their true conjunction.

This occurs when our fixed, natural Sulphur, through the astral Spirit or Azoth, is again resolved into a prima-terial water. This is then added to the Sulphur, through which the same is aroused anew to life, and through the subsequent New Birth, both are indivisibly united.

The water of the resolved Salt should not be added to the Sulphur or the Central-Fire all at once, otherwise, if too much water is added to the Sulphur, it would become totally dissolved again and would attain its earlier oiliness and it would float like an ordinary oil on the water, so that no unification of the Sulphur with the mercurial water could take place.

Here, in the artificial alchemistical Re-birth, the mercurial water must be weighed in such a manner and added to the Sulphur, so that it can digest the humidum radicale and change it into its nature.

That is how Sol and Luna contain themselves in their innermost essence and ignite in a new Centro of Life, until both have become an immortal and a lasting Being.

Since, through the sparse steeping of the azothic water, our Sulphur is nourished and exalted into a higher energy, that is why the Philosophers compare this with a young child which must be reared through milk and nutriments, until it has reached its mature age.

During this operation, great attention must be directed towards an even and strong administration of the external fire. Our foliated earth contains within itself a subtile, fiery smoke, a dry sulfuric fire, which coagulates the fleeting Spirit and gives it permanence.

The fiery smoke (hebr. Ruach), is located in the Centre of our Earth, and changes through its action everything fleeting into its fixed Nature. Then the movements of the Elements cease, which, when they again are dissolved with the fleeting, begin to move again, until the fixed has again overcome the fleeting. Then the movements cease again. This continues and repeats itself through solution and coagulation.

The external Fire should not surpass the internal. The internal fire is a mercurial sap, a nectar of heaven, which makes the materia alive, maintains it and nourishes it until the end of the work. That is how the fleeting spiritual moistness (our moist fire or Azoth) dissolves the fixed, earthly dryness (the fixed natural sulphur). The latter is again impregnated with the fleeting moistness; both intensify to a middle-substance (to a dry water), their sign is the blackness."

The Last Cooking.

The golden sap which we received after the exaltation of our stone, our elixir, is also called Lac virginis or Virgin Milk, it will now be placed into the philosophical egg (into a glass Phial) which is then hermetically sealed. Two-thirds of the receptacle must be empty so that the materia can circulate.

Then the well-luted phial is placed into an oven (Athanor), and the external fire is regulated in such a manner, that it does not overcome the internal fire since too great a heat would be destructive. If, on the other hand, the external heat is not sufficient, then the Spirit of the materia would remain motionless and idle, and it could not be constant and coagulate with the ground moisture.

The first work of our Magnum Opus occurred through the water and the earth, in accordance to the attributes of the external dark world. The second work occurred through the air and the salt; the third work will occur through the fire and the light.

When we know how to introduce the fixed, white Stone, the heavenly Light, as the dry Fire of the Philosophers, then we also know how to unite the white, fixed Queen with the red King.

That is why our Elixir is now being cooked in our dry Fire, until it becomes the completed red Lapis. Nevertheless, a continuous ignition of the dry fire, through the moist mercurial-Fire is necessary, through which the dry fire (our Sun-Son) must be nourished like through the blood of our heavenly pelican.

Even though our Sulphur and Salt were born again and resurrected through the impregnation of the Life-Spirit, they were not exalted in their powers, because the Spirit only associated with them, although at this point, did not permanently unite with them.

This unification occurs now through the coagulation and fixation of our mercurial Water and Sulphurs (or metallic Gold

Seeds) through the natural cooking in the eternal living Fire of the Philosophers. This Fire radiates and circulates through a spiritual Vapour ceaselessly upon our materia. It is warm, dry and moist; more spiritual than material. It warms our materia and cooks it gently through consistent radiations. It maintains, moistens, nourishes and multiplies it in strength and virtue.

This moist, steaming Fire (our Azoth) is in its nature like the materia of the Stone; it is taken of the purest substance of the same, and prepared through the art of the Alchemist.

This is the true Balneum of the Philosophers, however its preparation is a secret. It is the greatest Secretum of the hermetic art, only God can reveal it.

The cooking, which now follows, of all four Elements, must be brought forth one after the other in their order through the external fire, in seven gradations of cooking.

The first degree of cooking is governed by Mercury, because at this point, the materia is being made receptive and is brought to putrefaction through frequent steepings of our red Mercurii, namely, our fixed, natural Sulphur is again being resolved into his Materiam Primam Spermaticam or into a vitriolic Guhr.

The radical dissolution shows itself first through a yellow colour, which soon turns to deep blue, once the materia again settles on the bottom.

Following that, the materia again becomes deep black and becomes a Ravenshead, that means, a subtile muddy Earth, with which the reign of Saturn begins again. That is the second degree of the cooking.

Here the materia is called the philosophical Lead. It is considerably more exquisite than common Gold, because it is the materia prima for the tincture as both principles (Sol and Luna) are being united therein. Since the putrefaction is the main gradation of Alchemy, the tincture could not be prepared without the intimate unification of the male and female seeds.

As soon as the materia on the bottom of the glass solidifies and dries again and takes on the ice-grey colour, then Jupiter has conquered Saturn and has taken over the reign, in the third degree of cooking.

This is followed by seven cleansings or steepings, whereby the Spiritus volatilis is added seven times to the materia. When this occurs, the seven colours of the rainbow circulate (the so-called peacock's tail, cauda pavonis) up and down in the glass, but considerably faster and stronger, than it did in the second work. This is a sign, that Venus has begun its reign in the Kingdom of the Air, whereby the fourth degree of cooking has now been reached.

This Peacock's tail should only appear during the transition from black to white color. Should the Peacock's tail appear later, or should the black colour show up again, this is then a certain sign that something was incorrectly done insofar as the work was concerned and therefore, it was a failure. The same is true when a colour (such as the redness) appears before its time. The blackness should not show itself again at a later date, because once the raven is drowned, he should never return; yet the Dove should return often. That is what the Philosophers say.

Considerably more magnificent than in the second work, appears here the green colour, when it comes into view of the Alchemists in this last work, they say full of joy: "O blessed Green, without you, there would be no hope, of obtaining the Philosophers Stone!"

Finally, after a long and patient cooking, in the fifth degree of the cooking, the dazzling white colour shows up again. It announces the unification of Sol and Luna, which again form one body, the twofold, heavenly-made Mercurius duplicatus, our white Elixir of the second order.

This dazzling Whiteness is the second or the true Whitening of the Laton, the white Swan of the Alchemists, which swim in the vitreous ocean, the exquisite Albedo, the Head of the Lamb, our born-again Moon, the Sign of Fixation and the wonderful Marriage of the Sun and the Moon.

The fleeting Spirits became solid and this newly-risen, spiritual Body will be considered beautiful, white and immortal and it carries more than 100 different names in alchemistical Writings.

Now, the point in time has come, where the books can be burned.

This high White, our fixed Luna, which looks like a subtile Diamond-Dust in the Sun, is now ten times stronger in power and virtue than it was after the first whitening. The white Stone could be multiplied to such a degree, that finally, the white stone would become like a white oil, which would glow at night like the moon. Due to its magic attributes, the Philosophers call this the Elixir spiritum.

> "O! above the mercy of Jesus Christ towards us ungrateful and life's unworthy, we who do not want to acknowledge God's gift and grace. O! about my ingratitude, when I, O my Creator, through my pen and true, sincere and fatherly proposal, cause, that this our Queen would come into the hands of an unfaithful."
>
> – (Fridericus Gualdus, Philosophia Hermetica)

Nicholas Flamel said:

> "When you have reached the White, then you have cast down the raging bulls who, out of the nostrils, spew fire and flames. Now, Hercules cleansed the Augean Stables of its rottenness and stench, Jason gave the Dragon of Colchis the Juice from Medea, and you have the Horn of Plenty of the Goat Amalthea under your control, which showers you in your life with fame, honour and wealth."

Should you want to have the red tincture, then the materia must continue to be maintained in the fire. If it is removed, it would cool off immediately, and then it could not be brought to a red tincture. That is why the fire must be sustained, until the cooking enters into the sixth degree and with this, it enters into the Regiment of Mars.

Here, the materia has already become somewhat physical, so more heat can be applied.

Now, the materia becomes lemon-yellow. This colour appears to be higher and higher, until it finally becomes red-yellow.

Nobody can get from the black to the white colour, other than through the Peacock's tail, and nobody can get from the white to the red colour, other than through the yellow colour. The black stone represents the Winter of the philosophical work. The multi-coloured and the white Stone represents Spring, the yellow stone represents Summer, and the red stone represents Autumn, meaning the ripe fruit.

When the martial yellow colour has been attained, the materia cannot, through itself, come to a higher degree of redness unless you come to its aid with the secret red or solar Mercury.

The red-yellow colour eventually changes into a grey ash through further cooking, out of which suddenly bursts the magnificent golden-yellow orange colour that the Alchemists call the Phoenix. The seventh and last degree of cooking has been reached with this golden-yellow colour, and now begins the Regiment of the Sun.

With seven drops of his exquisite blood, our heavenly Pelican commences once more with his activity. Because now, with seven little steepings, the red solar Mercury, that means, the Virgin Milk of the Sun is being added, but only a little at one time, because this last operation is brought to its completion more by the dry Heaven's Fire.

Through these imbibitions with the solar Virgin's Milk, the materia prima is brought higher and higher to a redness, until eventually, the Colour of Perfection of the purple redness appears.

With this, the fixed gold or the red tincture of the second

order has been reached, which is ten times more potent that the one of the first order.

When the red Lion, our Son of the Sun, is left unopened in the glass for digestion, for two more months, he then becomes blood-richer, more beautiful and solid.

However, it does not coagulate into a hard stone, like glass or crystal, but attains the hardness of wax, which melts easily.

Eventually, when the stone remains rose-coloured, or ruby-red, then the work is complete.

Rosarium:

> "Born here is the King of Honour.
> No one may be Higher than him!"

When the Alchemist had finally prepared the red Powder, he exclaimed:

> "O, you holy Phoenix in my hands,
> you King of all Heavens,
> you Problem-Child of all Philosophers,
> you coagulated Heavens-Fire,
> now you are finally mine!"

Heinrich Khunrath:

> "When you have come this far already in this Life,
> you begin with the great Holiday of Holidays,
> and with unutterable Peace of Mind and Joy of the Soul,
> you dwell in the Spirit in the heavenly Jerusalem."

And the Alchemist Norton (a student of the Rosicrucian, Michael Maier) said this:

"On a certain day I overheard my teacher, when he said, that there are many patient and learned men, who with great effort and labour found the white Stone; but, in fifteen countries there might not be even one person, who is in possession of the red Lapis!"

"This is our King of the second Operation, a perfect Medicine for Men and Women. As an Elixir of the second Operation, it shows its Power, every ailment succumbs to its might. This is the Medicine for long Life; it prolongs it only through the Grace of God. Praised be the infinite, glorious Trinity, and the blessed, loving Virgin Mary.

With this Medicine, the all-mighty HERMES prolonged my Life without Herbs, the divine Majesty comforted me with the same Medicine and satisfied all my desires. United with this, until the end of the life, is a perfect treasure and absolute health. Pray to God with Love, for such a fortunate success.

However, our King lacks a crown. The third change will give him the crown; without this, his power resists the transformation of metals. The third change of our short way occurs through the fermentation as the accompaniment. Then he transforms himself into a red substance, through all the colours, as in the first and second work.

Three times the Elements must circulate in the fire resistant, united, mixed, putrefied, gloriously fixed, penetrating and cleansed, through which the Seed of Gold is being born again, and the power is also preserved indeed in actuality. This rotation must occur three times, so that the Medicine of the third order can be seen."

– (Fridericus Gualdus, Philosophia Hermetica)

"You should, however know, that from your Electrum a solution cannot be produced unless it has passed completely through the Circle of the seven Spheres three times, because this is the number which belongs to it, and it must bring it to its completion."

– (Paracelsus, Manuale de Lapide Philosophorum)

"Now, this Son of the Sun, this living, fixed and magnificent Gold, can bring the dead or metallic Gold to life, if this highly noble King is attired with his purple coat and again placed in his steam-bath, and remains there until he takes his coat off, then puts on a black coat, then a white coat and eventually, again a much more magnificent red or scarlet-coloured coat and by those means attains such strength, power and might, that from that point on, he rules over Death and Life, and he can give his subordinates that same might, magnificence and splendour, then he can also multiply his Kingdom without End."

– (The Sun from the East, 1783)

When the Lapis has passed through the third rotation in this manner and becomes a mineral Stone of the third Order, he will no longer coagulate into a powder. He will remain in the form of a ruby-red oil which, at night, throws off a wonderful shine and drives out all evil Spirits and ailments. It even can produce magic effects.

Because the Stone shines in the dark in this highly augmented form, the Alchemists call this a Carbuncle-Stone. However, from this point on, it cannot be multiplied any more because it could no longer be held by any Glass.

Remarks Regarding the After-Work.

"An also-renewed Spirit-Body comes into the Order of the Spirits or the exalted Bodies; they are no longer subject to the Laws of the coarse corporeity." — *(Ripley)*

It has been pointed out several times, when it comes to the Great Work of Alchemy, that we are dealing with the Mysterium of the threefold rebirth of the Spirit, the Soul and the Body. The white Stone of the second Order represents the perfect spiritual rebirth; the red Stone of the second Order represents the perfect astral (Soul) rebirth; and the red Lapis of the third order represents the completed physical rebirth. Hermes said the following about this:

"His power is perfect, when he (the fleeting Salt-Spirit) is transformed into Earth."

The whole hermetic work occurs in all three Worlds. The first work of dissolution, which ends with blackness, takes place in the lower, dark world. The second work of cleansing, which ends with whiteness, takes place in the so-called Astral (or Middle-world), and the third work of unification of Spirit and Soul takes place in the born-again Spirit-Body in the heavenly Light-World, the so-called mundus archetypus.

Two of these Worlds are considered to be non-existent by the Wise of the World. In spite of this, they do exist and they are highly real for every mystic with experience. The existence of the astral world can be proven through the dream-world; the existence of the dream-world cannot be denied. The pictures of the dream-sphere shine in their own light, the so-called Astral-Light. The mystics and quabbalists refer to it as the "Light of Nature." It is said, however, that dreams are shadows. This may be true in a certain respect, however where there is a shadow, there must also be a Light that produces and brings forth this shadow.

What we perceive from the so-called real Outer-World, are nothing else but pictures of our senses. The most important ability of the Spirit to interpret and retain these pictures is self-evident – it is the memory, and the ability to remember, that is, the visualization of past impressions repeated at random. Where are these pictures of our memory being stored? Some modern Psychologists consider the brain cells to be the seat of our perception and memory pictures. This hypothesis is no more intelligent than if someone explained that the molecules of the wiring in a radio are the seat of all the present and past sound-productions.

The visual memory is, without a doubt, stored in the deeper layers of the Soul, from where they rise to the surface of our consciousness, whether they are called or not.

Since the existence of an independent substantial Soul is denied by some modern academic psychologists, there is a mentioning of psychic occurrences in the sense of a heraclitic-actuality theory. All that can be said about that is, that a happening without a substrate seems to be something similar, like a river without water or an electric current without electricity.

Let us leave the Sophists with their "Psychology without Psyche."

The Soul substantially exists for everybody who does not consciously ignore it or, through sophism, allows it to be artificially disputed away. The Kingdom of the Soul is the inner World, which reveals itself through pictures and moods, which at first appear to be relatively unclear and chaotic. The reason for this is that the dream, or astral-body of a normal Human Being is not yet developed. Many are of the opinion that this inner life, this astral organism, would form by itself. But, nothing in Nature grows and takes shape by itself, if it is not given its nourishment and a corresponding model. Whosoever does not cultivate the Garden of the Soul wisely, but instead, allows it to grow wild, will never come out of the chaos.

The Necessity of Rebirth.

The Spirit and the Soul of the Human Being are only loosely connected with each other, and therefore form only a loose compositum.

The Spirit is the viewing, cognizant and the judging principle. Its activity consists of the formation of concepts, comparisons and conclusions. It is the principle of the consciousness, the reflection and the apperception, as well as the recollection.

In addition, the Spirit controls, with its will, the motorial nervous system of the body, which makes the conscious actions possible. Without the decisive consent of the Spirit, an act of volition cannot occur. That is why a Human Being is only responsible for those acts which he does with clear consciousness. Within this, lies the activity of the Spirit.

The Soul is the Principle of the (external and internal) perceptivity, the principle of the imagination, the perception and the receptivity. It is the plastic, creative, forming, producing and perceiving ability, which is creatively active as well as reproductively.

A Human Being can only remember that which, through perception, becomes conscious to the Spirit – everything else, he has forgotten. That means, it remains unrecognized and unconscious in the depth of the Soul. The Spirit can only become conscious of the content of the Soul, if he is closely connected to the Soul; that means, when the Spirit has immersed itself to a certain degree into the Soul.

As soon as the Spirit separates from the Soul, the Spirit can no longer remember the Soul. There are several conditions where such a split of the consciousness can occur to a more or lesser degree, for example: during sleep, in hypnosis, insanity, and in death.

Under normal circumstances, sleep is only a condition of

deepened fantasy, whereby the consciousness of the external reality is totally disconnected. Through that, the Pictures of our imagination in the dream appear to be totally alive and tangible – yes, they become one real reality. The Spirit can follow vaguely with his concepts and judgments, particular phases of the dream, but as a result of his greater loosening from the Soul, no longer has any more possibility of intervening to bring order into the course of events of the dream. He must endure these dream-images without will, because in this condition, he cannot turn towards the external reality, because for him it is disconnected, since the sensitive nervous system, which alone makes the sensory perception of the outer world possible, is not being controlled by the Spirit and Will, but by the Soul. The Spirit cannot refuse sense-impression, may they be by nature internal or external, but must acknowledge them, as long as he is connected with the Soul, if he wants to or not. Within this lies the passivity of the Spirit.

In a deep sleep, the Spirit is totally separated from the Soul. Upon awakening, he cannot remember anything of the dream, because in this condition, the Spirit cannot take on any sense-impressions, either internally or externally. Anaesthesia is nothing else but an artificially produced deep-sleep; a condition of deepest lethargy.

In hypnosis, an artificial separation of Spirit and Soul takes place, whereby contact with the outer world is being maintained for the present (= being aware of surroundings). When the hypnosis is being deepened, it then turns from sleep into a deep sleep. Also, the condition of hallucination and sleepwalking is based upon a separation of Soul and Spirit.

The hallucination is dreaming while awake, whereby an external reality is being ascribed to the pictures of the imagination. A particular subjective imagination was being projected to the outside, that means, hypostatized as an objective sensory perception.

The cause of this peculiar appearance lies in the partial segregation of the Soul from the Spirit. The Spirit somehow received through this an influence upon the sensorium and activates this in return.

During the condition of sleep-walking (Somnambulism), on the other hand, through a similar partial segregation of the Spirit from the Soul, the latter received an influence upon the motorial organism and through this activated, while dreaming unconsciously, the physical body. The Soul would not be able to do that in a waking state.

Insanity also shows, in its various forms, distinct characteristics of the segregation of Soul and Spirit, in that it appears as ailments. These are unnatural by nature, commencing with the lighter forms of manias (manic-depressive insanity), and obsessive or compulsive ideas, leading to graver forms of schizophrenia, semi-consciousness, or stupor. Schizophrenia especially shows all the characteristics of a split personality, whereby the transcendental Subject is divided into many personalities, which are mostly hostile to each other and eventually totally destroy the personality of the Human Being.

The ability of our consciousness to separate into two halves which act upon each other, cannot be exclusively limited to the dream-life. The cause of this segregation lies in the simultaneous existence of a waking consciousness, a dream-like subconsciousness and also a much deeper unconsciousness, which are separated from each other through two dividing thresholds (a threshold of consciousness and a threshold of perception). This cause is also given when awake.

All the content of the Inner World, which crosses over the threshold of consciousness, will become conscious to the Spirit, that is, the external empirical I. As soon as the I sinks below the threshold of consciousness, a partial segregation of Spirit and Soul takes place. The empirical I then becomes the transcendental Subject of the dream-condition; that means, the I of the Soul, and forgets, immersed in the picture world of dreams, the outer World and its earthly existence.

As soon as this Dream-I sinks to the lowest threshold of perception, it enters into a danger zone, where the personal and collective subconsciousness flow together, and out of the latter, dark chaotic figures emerge (encounter with the Guardian of the Threshold).

Should the Dream-I, the transcendental Subject of the Human Being, sink even lower than the threshold of perception, lower than the Limbus, then another segregation would take place, namely, the total segregation of the Spirit from the Soul. The transcendental subject would become a demon and would sink into the collective subconsciousness, into Hades, in which it would even lose the memory of the dream-world because it crossed the River Lethe. (Condition of Death and Madness)

As with the onset of sleep, a totally new world opens up (namely the picture-world of dreams, the kingdom of the personal subconsciousness). That is how it opens up to the Soul, which is segregated from the Spirit, with the onset of death an even much deeper world, namely the chaotic kingdom of the demons, the sphere of the collective subconsciousness, the underworld of antiquity, where the Soul has only a vague consciousness and a clear perception of the things and occurrences which surround her. The fear of death, which every normal Human Being has, relates in most instances less than the loss of the empirical I, otherwise, there would also be fear before going to sleep.

In death, when the Spirit and Soul do not remain inseparably together, then the Human Being enters into an abyssal World of darkness and demonism. When Human Beings are confronted with this, they become horror-stricken.

From this, you can see the urgent necessity of a rebirth of the Human Being, whereby Spirit, Soul and Body are joined together in an inseparable and immortal being, that even in death a separation of these three cannot take place.

The Three Worlds.

	LIGHT – WORLD, Ultra-mundane Heaven, Condition of Rapture. (2 Corinth. 12, 2-4)
Threshold of Recognition	
	Kingdom of the Spirit and Thinking. The conscious I – External World.
Threshold of Consciousness	
	Kingdom of the Soul and Feelings, Sensorium. Dream or Astral World.
Threshold of Perception	
	The Kingdom of Demons and the Depth-Soul. The dark Underworld – Hades.

"The Soul has, in the time of the outer body, three mirrors or eyes of all three worlds; this is the mirror she turns to, out of which she sees. But, by the Rights of Nature, she does not have more than one, that is fire-lightning, as the fourth figure in the dark world, in Loco, where the Principium originates, where the two inner worlds separate, one into darkness and the other into Light ... And when you place 1.) the light-world into its right fire, against as in primo principio. And 2.) the dark-world in the Fire-Root; And 3.) the external elemental World among the ordeal of the stars, between all that moves the great Mysterium of the Soul-fire. The Soul adjusts and devotes herself to this world, and she receives from this world the beginnings of her imagination. But, she has joined Adam in the Spirit of this World, and led her imagination therein, now her greatest desire lies in the ordeal of the Sun and the Stars and produces with it the Spirit of the external world, with the Being of the four Elements, always within her, and she has the greatest joy therein, where she is a guest in a strange lodge, because the abyss is below it, and there is great danger." — *(Jacob Boehme, Sex Puncta Theosophia)*

The Rebirth as the Mysterium of Initiation.

Even today, Plato's words are fully valid, when he said that those who sink in the mire, entered into the underworld uninitiated and unchastened, and only those who went through a mystical life, will enter into eternity. — *(Phaidon)*

According to Apuleius, the initiation is a voluntary death and it is the rebirth to a new and higher Life.

The endeavour towards true wisdom is comparable to dying, because the true Philosophers aspire to nothing more than the secession of the Soul from the earthly body. As long as the Soul is bound to the transitory body and is affected by it, the Soul cannot reach her true being, because the sensory perceptions are deceiving and nothing certain can be found in it.

The actual truth cannot be recognized by means of the body, but only through the Spirit, which has withdrawn from everything physical, the enlightened insight, the truth in us, that which is eternal and unchangeable.

The Spirit must concern itself with pure thoughts, which have withdrawn from everything earthly, that means, sensual. The one who totally immerses himself into the pure Spirit, lives solely in truth. For the true Philosopher, death does not contain any terror! Also, all the higher ethics must be based on the liberation of the Soul from everything earthly.

Only when the Soul is alone with herself, and she observes the pure, the eternal, the immortal and what always remains unchangeable in and through herself, then is she united with the wisdom. Then the Soul lives, as the Initiates say, in the Divine.

It is not possible for anyone to enter into the Society of the Divine, unless he has led a mystical life. He has to be totally free from everything earthly and transitory; such as only an Initiate, whose only endeavour is wisdom.

Those Souls, however, that are totally bound and incarcerated by the body are of a coarse earthly substance. Through this, they become coarsely material and impure. It is impossible for them to enter into the other world in a pure and clean condition, and they always pass away burdened with the earthly body, and that is why they do not reach the point, where they unite with the Divine, Pure and being One.

That Soul, however, that follows the higher insight and remains loyal to it, considers and nourishes herself always with the truth and the divine, and all that is superior to the intelligence-speculation, will be transformed into the nature of this divinity. She (the Soul) does not have to fear when separating from the body (while death occurs), that it enters into the chaos and that it will be torn apart there, and will be strewn into the winds and dissolved.

Even Human Beings, who are purely sensuous, have the spiritual-divine powers hidden within them, but only within the Initiate have they manifested in reality. Therein lies the transformation that took place in the Mystic during rebirth. The external world made a sensuous Human Being out of him, and he then was left to himself. Nature thereby fulfilled its mission. Nature cannot bring him to a higher perfection, than what is according to its own being, But wherever Nature stops, the art must begin. Therefore, Human Beings must take over by themselves, when it comes to this perfection. He must resurrect the God that is concealed in him by himself.

In antiquity, Human Beings were subjected to secret procedures in those places, where the mysteries took place. His lower earthly nature was killed in that manner, whereas the higher and divine within him was awakened to life.

The reason for the initiation lies in the confession, which were once given by Apuleius, who passed through the Isis-Mysteries himself:

"I was close to the Kingdom of Death, after I crossed over

the threshold of the Goddess of Death Persephoneia, I travelled through all the elements and returned again. In the middle of the night, I saw the Sun shining radiantly; I was close to the gods of the Underworld and the Upperworld, and I worshipped them from countenance to countenance."

And he adds to this, and rightfully so:

"See, I have reported something to you, which you do not understand yet, even though you have heard it!"

Much can be said about the Initiation into the Mysteries and about the rebirth, but to whosoever is not himself initiated and has not experienced the rebirth himself, these words are only sound and smoke.

Only the Initiated consider themselves to be authorized to speak about God and immortality. They know, whosoever has not passed through the Mysteries, and speaks about this, says something, that in principle he does not understand himself. Because, in the uninitiated, the Divine and the Eternal has not yet been awakened, and therefore, it does not exist alive. Would he speak of someone Divine, then he speaks about something which is Nothing. Only when the Mystic has carried out his trip through Hades and has awakened the Eternal in him, can he speak of God and rebirth and not until then.

What Apuleius wanted to say, was this:

"What I had in my mind was an infinite perspective, and at the end of that, was the Perfection of the Divinity located. In myself, I felt this power of the Divine. I carried to the grave, whatever this power kept down in me. I have died, when it comes to the earthly and was dead. As a lower Human Being, I had died. I was in the Underworld and I have travelled through the Kingdom of Death. There I associated with superior Spirits and I received from them important information or instructions.

In the middle of the darkness, I saw the eternal Light which illuminated my Path. After I had travelled through all the regions of the Underworld, I arose from the dead on the third day. I have overcome death, I became another, because I am transformed and I changed my form like Osiris, twelve times, whereby I passed through Earth, Water, Air and Fire. I have nothing to do with my former lower and transitory Nature because this, my Nature in me, is now saturated by the Divine. I belong now to those, who live eternally in the Land of the Living to the Right of Osiris.

I, myself, have become a true Osiris and hold the keys of Life and Death, and to dissolve and to bind the power in my hands."

The Initiation of the Mysteries of Antiquity is also an anticipation of those course of events and conditions, which occur at death. A mystical death, a philosophical dying, is just as serious and as real, as the true dying. Yes, it is even considerably more real, because it takes place with full consciousness."

The Osiris Mysterium.

The ancient egyptian Osiris Mysterium refers to the Inner, to the Soul of the Human Being, and to the course of events and transformations which take place there, and what determines the secret process of rebirth.

Osiris is the expression of the Soul's human drama. He is the pure human Soul, descending from the World-Soul (Isis). He is also a Son of the Sun, like Habal (Abel) is the Son of Adam-Kadmon.

Every Human Being is, according to the Soul, a part (Ray) of the original Osiris. However, within the Human Being, Osiris is dead, through the lower Nature (Set-Typhon), that means, through the evil demon, the animal Soul, which like Cain slew his brother Abel. He is buried in the shrine of the mummy of the Earth-Body and thrown into the Nile (River of Life).

But Isis (the Love for the Divine), is seeking, with the help of Anubis (reason), the parts of the corpse, to collect them and to care for them, that means, to mummify them. Out of that, the higher Soul (Horus), the new divine Osiris is born, which conquers the Demon (Seth).

That is how the Osiris-Mythus is the prototype of Initiation for Human Beings, who endeavour to awaken the eternal within themselves. The Human Being himself must become an Osiris, who experiences the fate of God upon himself. The Human Being who descends from the "Father" (Ex Deo nascimur), should produce within himself the divine "Son." Human Beings unknowingly carry within them, this Son of God as an archetype of the Soul, concealed within themselves. However, the eternal should become alive in them and reveal itself. But, for the time being this God, which is hidden in the Inner, is being suppressed in Human Beings, through the demonic force of the earthly Nature (Set-Satan). This lower, earthly Nature must be conquered and killed first (in Christo morimur), so that the higher divine Nature can rise up (per spiritum sanctum reviviscimus).

God (Osiris) is also killed and buried in the Human Being. The Human Being, however, must have him rise out of the mummy and bring him to resurrection. The fiery Soul is impalpable, incomprehensible and escapes into the heights of the free Ether, as soon as the Human Being grasps for her. He must, through the art of Alchemy, draw out of the Earth's volatile Sulphur of the Philosophers, this pure elementary Gold-essence of the Alchemist, somehow fix and bind it in a new manner, through an arcanum of the hermetic arts, so that it, although bound, yet be free and immortal. This Arcanum is the Mercury of the Philosophers, the mediator between God and the Human Beings, that is, between Spirit and Body. The Soul is fiery-volatile; the Mercury is a water-being, half volatile and half fixed. He can take on all the forms of the world, since he is extremely plastic; that is why he is also called Proteus. But, this Proteus must first be changed and tamed. He must shed his Dragon-Form and accept his divine Form before he can acquire the Gift of Prophesy.

God (Osiris) is dead and buried in Human Beings. The Human Being must, however, search for him in the darkness of his earthly being, like Isis (the Love of the Divine) with the Lamp of Anubis (the Light of Reason) searches at night for the corpse of Osiris, until she finally finds him in the waters of the Nile, that means, in the magic fluid, whose Life's Stream flows through and enlivens the whole earthly Nature (Egypt). However, Seth dissected the corpse (the Nature of his Being) of Osiris into 14 parts, that is, into 7 active and 7 passive spiritual-psychic potencies, as the ancient Sages of Egypt depicted them in their Hieroglyphs.

Isis (the divine Love) gathers and unites these parts and protects them with mummy-bandages of the preserving magic energies of her female principle, against putrefaction.

Eventually, Isis will be impregnated from the brought-to-life-again Osiris Mummy, and bears Horus, the new Osiris, God's Son, who can now say this about himself:

> "I am Yesterday, Today and Tomorrow;
> I am Geb – Osiris – Horus!"

That means: I am the eternal and unchangeable One, which was, is and will be.

Osiris was resurrected in the Son. The Son took on the form of the Father.

Geb is the egyptian God of the Earth (Saturn), that means, the God which is buried in the earthly body, the Father of Osiris, the Osiris of yesterday. Osiris himself is the out-of-Shrine, out-of-the-Mummy liberated, but in the water-dissolved Soul, the midnight sun of the Underworld in her course through the descending arc of the spiritual zodiac, wherefore the egyptian Priests whispered into the ears of Mystics during the Initiation, the secret word: "Osiris is a black God!"

The Mysteries of Osiris concluded with the discovery of the God. As Plutarch reports, the Mystics went, on the last night of the initiation, to the water, where the Priests carried with them in the holy drawer a golden receptacle. They poured water in this receptacle, whereupon the Mystics called out: "We found him, we wish good fortune!" Then, the fertile Earth was mixed with this water, and a human image was formed out of that, which was dressed and blessed with noble incense, to indicate that Osiris and Isis are Beings of the water and the earth. In paradise, Adam's body was also made out of water and earth!

Whereas Horus is the purpose of the Whole of Initiation. He is white, like Osiris is black. In him, as his Son, Osiris is resurrected. He is the one Out-of-Water and Spirit born-again divine Soul, the winged Sun, the divine Light-Falcon who rises up into the free ether and ascends to heaven.

That is the path of purification of the fiery Soul through Earth, Water and Air, until she will be again united with the divine Light, and can say about herself:

"I am Yesterday, Today and Tomorrow,
I am Geb – Osiris – Horus!"

The Mysteries of Eleusis.

As the grand drama of the Soul in the egyptian mysteries took place, so is how it happened in the greek Mysteries.

There were a number of minor mysteries and number of great mysteries, into which the candidates would only gain entrance by degree, and only after a strict previous examination.

The minor Mysteries were held every year at Agrai. There were the preparations for the great Mysteries, and they were held every 5 years in the Autumn at Eleusis. Any disclosure about the Mysteries was punishable by death. New members were called Neophytes.

The minor initiations began with preparatory symbolic purifications, a bath in the river Jlissos. The Neophytes also had to sacrifice a pig, and fast for several days.

Then they would be led by Hierokeryx (the Herold of Holy Things) into the Temple region of Demeter. Here a choir sang an ancient dorian Song about life as a dream and the real Life.

Then the Neophytes were led by Hierokeryx (which represented Hermes), into the sacred grove, where the climax of the minor Mysteries took place; the first act of the sacred drama of Persephone, which was presented to them live, in the form of a drama.

In an utmost impressive and breathtaking scene, it was shown to the Neophytes, how Persephone (the Human Soul), succumbed to the temptation of Eros (the demon of the lower Nature), instead of thinking about Dionysos, her Husband appointed from heaven (the divine Light), and being subsequently abducted by Pluto (the Earth-Zeus or Saturn) into the Underworld, to be imprisoned there forever (the captivated Soul in the bounds of the coarse corporeality).

After Act 1 of Persephone's drama, the Neophytes were deeply moved by the performance. At this point, it was explained to them by Hierokeryx, that they had just witnessed the history of their own human Soul.

It was further explained to them, that even now, they were walking in darkness, in a life that was nothing more than an appearance. At one time, however, they did live the real true life, until they were blinded or bedazzled through the sorcery of Eros and fell into the earthly abyss. They were advised to reflect upon the words of Empedocles, who wrote:

> "that the formation of a Human Being was an awful catastrophe, through which the eternal, living beings became a mortal."

Now the Neophytes had become Mystics, that means, veiled.

They recognized that their present life was only a transition towards their true existence. They are the "Veiled," because they had not seen the "Great Light," the full truth, yet, but they suspected it, and they saw it from afar, like through a veil.

They also had to impress upon their minds, the statement made by Olympiodorus: The Purpose of the Mysteries is to bring the Souls back into the condition, from where they (before the fall into the lower world) originated.

Prepared in this manner, the Mystics expected further enlightenment about the mysterious process of rebirth from the experience of the great mysteries, through which they are led to the "Great Light" and in this manner become true initiates, intellectuals, seers (Epopts).

The Great Mysteries of Eleusis.

Only the Mystics were allowed to partake in the great Mysteries; those who had already received the minor initiation. The great mysteries took place every 5 years, at the time of the harvest in the month Boedromion (September) in Eleusis for the duration of 9 days.

The First Day was called "The Assembly." The mystics gathered for admittance to the great Mysteries and prepared for it.

The Second Day was called "To the Ocean, You Initiates." On this day, they walked in a festive procession to the ocean's shore to undergo a symbolic purification in the ocean water.

On this day, the showing of the sacred Drama was continued. Act 2 was shown, the subject of which was the pain and despair of Demeter (World-Soul) over the loss of her daughter Persephone (human Soul). It was portrayed to them in a moving manner. They saw how the grieving Demeter, for nine days and nights, wandered about with a torch, searching for her lost daughter, as Isis once searched for the missing Osiris. Hierokeryx explained to them the esoteric sense of the sacred drama. He spoke to them about the divine Love of the World-Soul, which is searching for the Soul of the Human Being, to liberate her out of the bounds of mortality, and to unite herself again with her. Then he spoke to them about the transformation to which the human soul, during her migration through the different levels of existence, is subject. Holy Songs concluded these celebrations.

The Third Day was "The Day of Mourning." On this day, the Mystics, while mourning with Demeter for Persephone who was staying in Hades, were reflecting upon their own souls. Esoteric Exercises, such as holding communion with oneself, fasting, prayers and quiet meditations were taking place. In the evening, the Mystics gathered for a sacred meal. At the conclusion, they drank a mysterious mixed drink.

On the Fourth Day, offerings were given to Demeter and Persephone.

The Fifth Day was "Torch Day." At the onset of darkness, the Mystics were led by Daduchos the Torch-Bearer (representing the Sun), with burning torches to the Temple of Demeter, to symbolically repeat the search for the Goddess. The burning Torch represents the symbol of the Love ignited from within for the Divine.

The Sixth Day is called "Iacchos." It was the most festive and the highlight of the Great Mysteries.

On this day, a column of Iacchos-Dionysos was carried in a festive procession by the Initiates and Mystics for over four hours on the long, sacred road from Athens to the Temple of Demeter in Eleusis. This was done to symbolize the introduction of the Light and the presence of the Divine. Every Mystic received, before the beginning of the procession, a Thyrsus-Staff and a sealed basket, the sacred chest which they had to carry with them the whole day, and were not allowed to open. The sacred chest contained three blessed objects, and was opened in the forthcoming grand and sacred night of the initiation by the Hierophant himself, and he showed and explained these things to the Mystics. The Hierokeryx also told them that even the carrying of the sealed basket had a deeper meaning for them. Because, even as a Human Being, they carried all kinds of things with them, of which they had no knowledge; concealed abilities that would only reveal themselves at a later date.

At the onset of darkness, the Mystics were received in the Temple of Eleusis by the Holy Herold, with the call: "Escato bebeloi!" – "Begone from here, all unconsecrated and ungodly, whose Souls are covered with crimes!," and drove thusly all of the trespassers out. Unauthorized intrusion to these secret celebrations were punishable by death.

The Mystics had to cleanse themselves once more in blessed

water, and took vows not to disclose anything to the uninitiated of what they would now see and experience.

After the vows, Hierokeryx told the Mystics that they were now on the threshold to Persephone's infernal dwelling. However, in order to reach the "Great Light," they had to walk first through the darkness. In order to understand something about the true existence of their Soul, they must first walk through the Kingdom of Death. This is the test through which they would change from Mystics into Initiates (Epopts). The Mystics now had to undress and they don a deer hide (the Symbol of animalism, the "Coat of Skins," which Adam received upon leaving Paradise).

Then Daduchos extinguished his torch. All other torches were also extinguished, as in Death the Light of the external World becomes extinct.

Now the Mystics are led by their Mystagogues to the entrance of the infernal labyrinth. Total darkness reigns there. This represented the condition of their Soul, which only possessed her illuminated natural intellect from the external sensual Light, which is being filled with the deepest darkness, as soon as the external sensual Light becomes extinct. Such a Soul knows nothing about her once-higher existence, because the "Great Light" of the true Recognition and Initiation has not yet risen up in her.

The Mystics moved forward slowly and with uncertainty in the dark labyrinth. They were being transposed into a particular mood through fasting, prayers and instructions, and through a mysterious mixed-drink. Eerie noises, horrifying moans, and terrible screams reached their ears. The roar of thunderbolts, the force of which shook the arched passageways, startled them with fear. Glaring lightning pierced the darkness and showed the frightened Mystics gruesome appearances, so that they were overcome with dizziness and were horror-stricken. But only for moments did they see this terror that surrounded them; then they were again enveloped by absolute night. The Mystics also felt

131

themselves being seized, beaten and thrown to the ground by invisible hands. Fearful natures preferred at times to turn back through the labyrinth to seek the exit. But, by doing that, they lost forever the Right to receive the initiation.

Even though the Mystics were still in their physical bodies, in this hour however, through the art of the Priests, the curtain that separates the invisible world from the visible world was lifted, and they were granted access to the lower demonic depths of the Spirit-World. Plutarch, who was initiated into the Mysteries, compares the horror that the Mystics experienced in the infernal labyrinth, with the terror of death.

Finally, the Mystics saw a dim light in a crypt. It illuminated a horrible scene. They saw the Tartarus, the brazen gate opened with a horrible noise. They saw the domicile of the from-the-furies tormented damned?, they heard their lamentation, their fruitless repentance, the sound of fear and the longing for the lost paradise. In between the voice of Hierokeryx could be heard, who at times shouted to the Mystics words of warning and then threats, and at times gave certain explanations of what represented itself to their eyes and ears. Then the Gates of Hell closed and the Herold communicated to the Mystics, that they would now enter the Plutonium, the domicile of the Ruler of the Underworld, where they would view the Third Act of the sacred drama.

Amongst solemn festive singing by invisible Spirit-Choirs, the Mystics entered now into a large, endless, illuminated, infernal hall, which was filled with an eerie twilight.

The ceiling was carried by an out-of-copper embossed, dull, shadowy elm, the "Tree of Dreams," whose pale silvery leaves roofed the whole room. From up above, through the branches, giant bats and grimaces stared down upon the Mystics.

Pluto was sitting on a black ebony throne, with a Crown of thorns upon his head, and a sceptre in his hand that was double-pointed at one end. Next to him sat Persephone. Covered in a black veil, her face reflected great sorrow.

The Hierokeryx now explained to the Mystics that they should recognize in Persephone's destiny, the drama of their own Souls. Just as Persephone suffers under the rulership of Pluto and is longing for her mother and her eternal Light-Homeland, that is also how their soul is suffering under the might of darkness and sensuality, and is longing unceasingly for the heavenly Light. The grimaces that stare at them out of the "Tree of Dreams" are phantoms of the past; transitory pleasures and suffering, which prevent Human Beings during their earthly life, which is only a dream, from the real life. The Herold kept silent. Persephone however, was giving her suffering and longing, a heart-wrenching expression. The Mystics had to make a Flower-Offering of Narcissus-Wreaths.

All of a sudden a big double gate opened, and a radiant light penetrated into Pluto's gloomy Hall. A shout rings out: "Come you Mystics, come here; Iacchos-Dionysos is here! Demeter expected Persephone! Evohe!"

Persephone jumped up, as if she were awakened from of a deep sleep. "Light," she cried out, "my Mother! Dionysos!" She wants to rush off towards the Light, but Pluto grasps her and forces her back into her seat. There she collapses and dies.

With one stroke, all light becomes extinct, and in the deepest of darkness a voice speaks: "Dying is to be reborn again!"

What is spoken of here is the mystical death. Since Persephone (the Soul) dies to the Underworld, she becomes free of the might of Pluto (the Earth) and returns to the true existence in the divine Light.

Now the Mystics were led to the surface again by the Mystagogues. The terror of the Underworld and the Tartarus is now behind them. On the surface, they are being received again by Daduchos and Hierokeryx. They have to take off their deer hides and bathe in blessed water. They received white raiments and are led into a gigantic Temple, which sparkles in the brilliance of a thousand torches. Now they are being led to the Hierophant, who is dressed in purple. He reads to the Mystics, from old stone plates, regarding the hermetic secrets of the process of rebirth; revealing these secrets carries the death penalty.

The sealed baskets were now brought in and given to the Mystics, and then the Hierophant opened them. After the Mystics removed the objects (One Egg, One Pinenut, and a spiral Snake made out of copper) from the basket, the Hierophant explained their symbolic meaning. The egg, with its seven components, is like the seven colours of Persephone's veil, a symbol of the seven spiritual-psychic potencies of the nature of a Human Being, which was already known to the old Egyptians.

The Pinenut is a symbol of the divine Light-Seed, which is concealed in it. During the sacred procession, they already carried the Thyrsus-Staff with the pine cone. They should remember that the Titan-Son, Prometheus, who formed the first Human Being out of moist clay and brought him to life through the divine fire, which he brought from Heaven, hidden in a Narthex-Staff. This divine spark is not brought to life in everybody. That is why Plato said:

> "There are however, many Thyrsus-Carrier, but there are only a few enthusiasts" ("polloi men narthekophoroi, pauroi de te bakchoi" – that means, many are called, but only a few are chosen).

The spiral Snake is composed of the metal of the foam-born Aphrodite; she stepped in Cyprus on land. Aphrodite is the Love, which entwines Persephone (the Soul) like a snake. However, this snake must wind itself upward, instead of winding itself downward, to become the snake of wisdom, as the two snakes on Hermes Staff wind themselves upwards, up to the golden-winged sun. (The Uraeus – the snake on the forehead of the Egyptian, as a symbol of divine wisdom).

As soon as the Hierophant has finished, the door in the Temple to the inner sanctum opened, and radiated in the brightest of lights. Heavenly sounds enchanted the minds of the mystics. Wonderful scents seasoned the air. In the distance, they could see lovely groves filled with eternal spring and meadows of the Elysium covered with flowers, where the departed were peacefully resting or strolling.

Now they were also viewing the fourth and last act of the sacred drama. They viewed how Persephone (the human Soul) was liberated through her mystical death from the power of Pluto (the earthly corporeality) and then led by Hermes (the higher Genius) and returns to her paradisiacal homeland, where she is brought back by her mother, Demeter (the World-Soul), and by Dionysos (the Divine Light), to her father Zeus (God-Father).

Following that, a hymn was sung to honour Zeus, Demeter, Dionysos and Persephone. Then the Hierophant bestowed upon the mystics the highest of blessings, with the following words:

"May your wishes be fulfilled, return back o human Soul to the Soul of the World!"

The holy initiation was now concluded, and the mystics had become seers, epopts, true initiates. A happiness never before known to them, and an inexpressible peace filled their hearts. They had overcome the fear of death, the dark puzzle of life was solved, the "Great Light" of Eleusis had arisen in them.

The Effect of the Grace of the Mysteries is praised by a homeric hymn to Demeter, with the words:

"Blessed of the mortal earthly dwellers are only those, who have seen this; those however, who died without this initiation and if they were strangers to this holiness, they do not have part of this share, instead they are dead in the moist mess of the Kingdom of the Night."

Greece's greatest lyric poet, Pindar, glorifies the Eleusinian Mysteries in the same manner:

"Blessed, who has seen them and then descends below the Earth! Only he knows life's goal, he knows also the God bestowed beginning."

The Tragedian Sophocles said the following about the Eleusinian Mysteries:

"O blessed, thrice blessed are those mortals, who have at one time seen these initiations and then descended to Hades; for them Life is only there, the others await only hardship and severe misery!"

The Eleusinian Mysteries flourished from 150 B.C. to the end of the 4th century A.D.

It was a grandiose system for conducting Souls, it was an extraordinary blessing, it had a character-forming effect, and it made the Greeks into a Nation of Philosophers. In the times that followed, up to now, there have been no institutions, which can be compared or even come close in spiritual depth, religious seriousness and psychological mastery to the Eleusinian Mysteries.

Virgil and Dante attempted poetically to sympathize with the Eleusinian Mysteries, but, through the mere word, they were never able to replace the immediate fascinating power of the liveliness of the sacred drama.

It was one of the darkest hours of mankind, when the zeolitic Emperor, Theodisuis the Great, upon the instigation of the Catholic Church closed the Temple in Eleusis in the year 385. Before that, he had the Alexandrian Library burned, by orders of the same authority, and thusly robbed mankind of an irreplaceable treasure of knowledge and culture.

The Soul of a Human Being is placed between the Materia (Body) and the highest of intelligence (her immortal Spirit or Nous). Which of the two will be victorious?

The outcome of a conflict of a long life lies within this triad. Should the Soul, during this time, be predominantly occupied with purely physical pleasures, then the destruction of the earthly body would result inevitably in the death of the astral body. Through this, the astral body would be prevented to unite with the highest Spirit of the triad, which (the highest Spirit) alone can transfer to us, immortality.   – *(Galatians 6, 8)*

Or we become immortal Mystics before the death of our

136

earthly body and we will be initiated in the divine truth of the Life hereafter. That was the main subject of the Mysteries, and was considered by the catholic Church to be satanic, and regarded by modern Psychologists as being ridiculous.

However, to deny that Human Beings are in possession of certain concealed powers that can be aroused through the hermetic art, that Human Beings can unfold to the highest degree, that Human Beings can become initiates (even a Hierophant) and can communicate this knowledge to others is a gross insult to a great number of the most outstanding and wisest Human Beings of antiquity and the middle ages.

What the initiate was allowed to see in the last hour of the holy mystery, can only be cautiously implied. Philosophers such as Pythagoras, Plato, Plotinus, Iamblichus, Proclus, Plutarch had knowledge of this, and they and many others confirmed its reality. Whatever the last Epopteia was, was implied by Plato in Phaedrus: "... after we had been initiated into these mysteries, we were liberated from the annoyances through evil, which would have awaited us otherwise with certainty in a future period of time. Likewise, consequently of this divine Initiation we became onlookers of a totally homogeneous, lasting and merciful visions, which dwell in the pure light."

Out of this, you can gather that the most exalted of the Epopsy, consisted in that the gods themselves could be seen in their own light. That was not a pathological hallucination, as modern Psychologists would like to make us believe, but it is the awakening of a totally new sense; a supernatural Spirit-Eye, with which the initiate can see into the World Beyond, and in this manner can, from the Beings which exist there, and from the course of events that take place there and conditions that exist there, procure unfailing knowledge through their own immediate perception.   – (2. Corinth. 12, 2-4)

The Mysteries of Israel.

Everything in the holy Torah (Bible) has, in accordance with the mysterious trinity of human Nature, a threefold sense; firstly, the external historical meaning of the word (Paschut), which corresponds to the body and the forecourt in the temple; secondly, the allegorical-mystical sense (D'rasch), which corresponds to the Soul and to the Saints in the temple; and thirdly, the inner secret sense (Sod), which represents the Spirit and the Holy of Holies.

The external, historical meaning of the word is like everything external in the World; only an appearance or symbol. It is so often contrary to reason, that no reasonable Human Being can take it for the truth. Yes, the irrationality should be a protection against the literal word-for-word understanding of the Holy Scriptures.

The "History" of Israel, which is quoted in the Torah, is therefore not an explanation of the external historical events, but in truth, it is in accordance with secret quabbalistical Midraschim, the allegorical History of the human Soul and her mystical rebirth.

Just as once Israel departed from Egypt (the Land of Slavery), to search for the Promised Land, the Soul should also depart from the external earthly world (Egypt), in which she was imprisoned in the bondage of the senses and lower desires, and search for the lost paradise.

The Children of Israel (that means, the Souls struggling with God) are being led by Moses (Thoth-Hermes or Mercurius), who with his Aesculapius-Staff (the Power of Wisdom), parted the Red Sea (Chaos).

The Children of Israel (Souls) now walk dry through the parted Sea, which is therefore a dry water, which does not wet the hands, while the Souls led by the divine genius pass through the chaos without hindrance. However, the Egyptian (worldly Human Beings) drown in it.

It is the inner Chaos, through which the Soul must pass; the inner sea of malice and bitterness, the Kingdom of Demons, whose restless floods recede and are being banished by the will-power of the Soul, which is being led by divine Light.

All those who leave Egypt (the world) will encounter such a sea, in order to reach the Promised Land. However, the true Hebrews can pass through it unmolested, that means, the true crossing over Souls. Because hebraeus means: the one from across.

> "I speak as concerning reproach, as though we had been weak. Howbeit whereinsoever any is bold, I am bold also: Are they Hebrews? So am I," said the Apostle Paul and at the same time, he mentioned how he spent a day and night in the depth of the ocean! — *(2. Corinth. 11. 21-25)*

As soon as the Soul turns away from the external World, she begins to become a dweller of the desert. When she returns out of Egypt's Darkness (= Worldly Errors) to the secret place of the Heart, she will not find anything else, but a place of terror and an uninhabited wilderness. Because, that is the desert, impassable and waterless land, the long neglected conscience, totally undeveloped, covered with spines and thorns, full of terror.

The soul must leave Egypt and search for the desert, and the following is being said about it:

> "They shall build the old wastes, they shall raise up the former desolations, and they shall repair the waste cities, the desolations of many generations." — *(Isaiah 61, 4)*

In other words: the Soul must leave the external World, and must return to itself, into her inner self, to search there for the entrance of the lost paradise.

That is a Journey and Change which requires many years, and many times during this journey through the desert, the desponded Soul will ask:

> "Is the Lord among us, or not?" — *(Exodus 17,7)*

Yet, she will receive nourishment from Heaven and oil out of the rock, the dew of Heaven and the fat of the Earth. She will ascend to GOD's mountain (being initiated) and GOD will place once more His Commandments into her heart. Even though, the Soul will be still pursued by the snakes of malice and temptation, Moses (the divine Genius) will teach them how the snake is exalted. Yes, the Soul might even again, worship the golden calf, into which Isis once placed the dismembered parts of Osiris. But Moses (the Divine Spirit) will burn the golden calf of Isis (the astral Gold-Corpus) with the Fire of the Philosophers and dissolve it into a red dust, out of which the Hermetics prepare their aurum potible.

Then the Soul will again erect in the desert, the true "God's Hut," with its forecourt (the body), where upon the altar, the burnt offering of a calf (animal Soul) is being sacrificed; her sanctuary (the Soul) with the water of purification, the incense-alter of the sacred prayer, the seven candlesticks of her inner Human Being, which must be ignited by the Divine Light, the holy oil or the thick water of the Maccabees, so that the divine spark does not become extinguished, and the table with bread and wine (the heavenly nourishment of the inner life). Between the sanctuary of the Soul and the holiest of the holy of the Spirit is located the curtain of reason, only through this can the Soul reach the innermost.

The sanctum sanctorum contains the golden ark of the covenant or the mystical shrine of the Law (the spiritual body of the WORD OF GOD, that means, the LOGOS).

Finally the Soul reaches the end of the Desert-Journey or the inner communion to the heavenly Jordan (JORDANUS MAXIMUS = the water of life of the Nazarene). Here the second crossing takes place.

The first crossing of the Soul occurred from the World into its own inner. The second crossing occurs when the Soul passes through itself and enters into GOD.

During the first crossing, the sea flees; during the second crossing, the Jordan turns upwards. The flight of the sea means the destruction of the bitterness. The turning back of the Jordan is the reinstatement of Love. Because the JORDANUS MAXIMUS, the Nazarene, is the Water of Grace, which is only bestowed upon a humble Soul. Jordan means: Descend. It is a voluntary humbleness and humiliation.

The first crossing took place through the denial of the world. However, in the second crossing, a Human Being denies himself and places himself into the utmost humbleness. He who recognizes and leaves his own evil (his lower demonic nature), has consummated his first crossing. Whosoever does not consider his Goodness (his virtues) as his own, but attributes them alone to God, has carried out the second crossing and renounces even his own I. Then the second miracle occurs and the Jordan turns upwards. Then the divine grace sets in again and exalts the Human Being to the Son of God, who does not do anything by himself, but who only carries out the Will of the FATHER. The Jordan of the own Soul has been passed through, Paradise has been found again, the Human Being once again is united with GOD.

The Christ – Mysterium as the
History of Initiation.

What was said about the threefold sense of the sacred Torah, applies in the same manner to the early christian Gospel, that is, the Life-History of Jesus.

Jesus (the higher Reason-Soul) is born in a stable of the animal-soul by the Virgin Mary (the Love for the Divine). His FATHER is the eternal Light, which is enthroned above the Heavens. Through the baptism in the Jordan (the voluntary "Descent"). He receives the holy Spirit of Wisdom, and becomes in that manner a "Son of the Light."

Then He is led by the divine Spirit into the (inner) desert, where He is for forty days and nights "with the animals" and struggles with the Demon, until the Angel (the divine Genius) is victorious in Him.

Now He can change the Water of the Philosophers into a philosophical wine, just like the vine (which He is Himself), that draws the water out of the Earth and changes it into wine (Spiritus).

Henceforth, the rebirth of a Human Being out of Water and Spirit (Spiritus) is the main-content of His teachings.

The enlightened Reason-Soul now heals everything which is ill in us, removes our spiritual blindness and deafness, and drives the devils (demons) out of our inner.

However, the Pharisees and Scribes (the lower reasoning or intellectual powers), who are blinded by the demons, seek to kill Him and try to prevent Him from doing His Work, but, with their pseudo-logic, they are not able to extinguish the Light of Wisdom.

Jesus (the enlightened Reason-Soul) knows that He cannot become immortal in the Earthly Body, that is why He takes His cross (the lower quadruplicity) voluntarily, to have it totally die off.

The voluntary Self-Crucification is the foundation of the whole rebirth, that is why Jesus said to His disciples:

"If any man will come after me in rebirth, let him deny himself and take up his cross!" — *(Luke 9-23)*

Now the last struggles begin, because he had provoked his greatest enemy (the lower Nature). The Gethsemane Struggle begins. While doing so, bloody Sweat is coming out of His body (the extracted Lionsblood!). But, the Angel (Divine Genius) strengthened Him with the Water of Life.

But once more, He was under the influence of the Power of Darkness. The Tetrarch Herodes (the Prince of the Lower quadruplicity) would have liked to use His Wisdom for his own dark purposes. Since however, he did not receive from Jesus (the enlightened Soul) any explanations, then he eventually clothed her (the enlightened Soul) in a white garment to ridicule her, and sent Him to Pilate, the governor (World-Spirit), who also does not know what to do with the enlightened Reason. However, since she spoke to him about the true Kingdom of the Light, the World-Spirit or the lower Reason (Pilate), did not recognize the truth, therefore, he dressed the intelligible Soul in a purple coat to ridicule her.

Jesus (the enlightened Reason-Soul) knows, that she first has to enter into the blackness, before she can attain the miracle of the true whiteness and redness.

Jesus (the enlightened Soul) is now being crucified on the cross and was given vinegar (= Azoth) to drink. Now the mystical death sets in with its total might. A terrible tear ripped the curtain of His inner temple. The spiritual Sun looses its shine, and deep darkness envelopes Him. The Sun and the Moon are separated, and the soul cried out in this, the most extreme darkness, the gruesome "ELI, ELI lama asabthani! My GOD, my SUN, why hast thou forsaken me?"

Now with the philosophical spear, the side of the body is pierced, so that the red leonine blood and the white Gluten of the Eagle comes forth, which the Soul takes along with her. The lower quadruplicity is now divided, and now the Soul can begin her journey to Hades, to purify and coagulate again those two tinctures.

The separation of the Soul from the Spirit, the Moon from the Sun is however necessary, so that the Soul, on her journey through the different spheres of the underworld, can dissolve her old astral image or corpus and can bring it to a spiritual rebirth.

– (Hebrew 4, 13 / Jeremiah 23, 29)

But on the third day, the Soul will rise from the dead again. As it is written by the Prophet:

> "Come let us return unto the Lord: for he hath torn, and he will heal us; he hath smitten, and he will bind us up. After two days he will revive us; in the third day he will raise us up, and shall live in his sight."   *– (Hosea 6, 1-2)*

Then the Morning-Redness (dawn) will break, and the Soul will be again united in a new Spirit-Body with the Light forever, and ascend to Heaven to her divine FATHER.

The Two Paths.

The Soul of a Human Being is situated at a crossroad, like Hercules. She has to chose between the sensual and transitory outside world, or the spiritual and eternal inner world.

If she chooses the latter, then everything that a Human Being is attached to in his everyday life, must lose all value to him. New values must replace the old ones; a total transformation of the emotional and spiritual life. If, up to now, the outside world represented the only real world and the inner world was nothing but a shadow of the outer sensuous reality, the whole relationship to the world now turns around. Only the things of psychic-spiritual life are real; the outer sensuous world, by comparison, is not real.

For the one, however, who has changed his thinking and feeling towards the external world to such a degree, the external world will lose its absolute value. You will, of course, see this world, but in an entirely different light. The inner world will also appear to be totally new; but you also must learn to feel and be sensitive to what you see. Whosoever is inclined towards sensuality with active feelings and sensitivity, looks upon the creations in the inner world as mere creations of fantasy. He does not live in them, therefore, they are mere images, that is, reproductions of the outer sensual or material world.

Whosoever has withdrawn his feelings and his sensitivity from the outer world and directed them towards his innermost, has gained at the beginning of the mystical path, the right to an experience which deeply moves his Soul.

It is the experience, that the Human Being is losing his emotions and feelings for the external world, but the new world does not open up for him. The old values are gone, but no new values have come into being within him. For whoever steps onto the mystical path, at one time, that will become realty. Now the Human Being has reached the point, where the Spirit interprets

145

all life as death to him. He then is no longer in the world. He is a dweller of the desert, he is under the world, in the underworld. He is on his journey in Hades.

Blessed is he, who does not sink into chaos, but where instead a new world opens up for him. He either fades away in this dark abyss of the Gnostics, or he rises again as a transformant from the dead. Then before him stands a new Heaven and a new Earth. Out of the spiritual fire of the Sages, to him the whole world is born again.

Everything is flowing. In the material external world, there is only coming into being and passing away, no existence. Nothing remains, nothing is permanent in this world of continual comings and goings. True permanence can only be found by looking back into the past and by looking forward into the future. This is worthwhile to be found. This is Divine; that what was, that what is, and that what will be. That means what is "ONE." Because in GOD, the past and the future are united into eternal presence:

"With whom there is no variableness, nor change of light, or darkness."   – (James 1-17)

God does not change Himself. He is always the same, semper idem. Only Human Beings change. They are never the same, but always at odds, always divided within themselves.

GOD is the centre of the All (Universe); He is also the centre of our Soul. The Souls of the Gods move steadily around this centre of the universe, and since they move around this centre, that is why they are Gods. A God is the one who is connected with this centre. But since a part of our Soul is enveloped by the body, we are only with the part which is not immersed, connected, through our centre to the centre of the All (Universe). We do not always look upon this secret centre of our heart, but when we look upon it, we have our goal and our peace in sight, and, because the Soul here, is in the body, there is falling away, and flight, and loss of the wings. Plutarch said:

146

"A Human Being is put together; and those who think that a Human Being consists of two parts, are in error. Those who think that, are of the opinion that intelligence is a part of the Soul, those who think that, do not err less, than those who think that the Soul is part of the body; because the intelligence (nous) surpasses the Soul by as much, as the Soul is superior and more divine than the body.

Reason comes into being through the connection of the Soul (psyche) with the Intelligence (nous), and out of the connection of the Soul with the body comes the passion (thymos), which is the beginning (or the principle) of pleasure and sorrow, and the other of virtue and depravity. By these three parts being closely linked together and condensed, mankind was given the body by the Earth; the moon gave the Soul, and the sun gave the Intelligence.

Now, in regards to the deaths, which we have to die of, one will make the Human Being out of a three a two, and the other, out of a two, a one. The former is in the region and under the righteousness of Demeter. That is why the name that was given to the mysteries (telu), was like the one given to death (teleute). The Athenians called the deceased those "who are dedicated to Demeter.""

In regards to the other death, it is in the moon or in the region of Persephone. With one, dwells the earthly Hermes, and with the other, the heavenly Hermes. He gathers the Soul suddenly and with force from the body; but Proserpina gently wrenches the Soul from the body, and it takes a longer period of time. For this reason, she is called Monogenesis, or the descent of the whole human race from a single pair, or asexual reproduction, because the better part of the Human Being will be by itself, if it is separated from her.

In accordance with Nature, one thing will happen as well as the other. It was decreed by Belief, that the Soul, may she be with or without Intelligence (nous) when she has escaped the body, for a while (not all Souls remain the same amount of time) must travel in the region which extends between the Earth and the Moon. Those who were unjust and wanton suffer their punishment here in accordance to their evil deeds. But the Good and Virtuous are kept here until they are purified and have, through perspiration, discharged all their contagious germs which they took with them through contact with the body (as it happens through putrefaction), living in the mildest region of the air, called "The shadows of Hades," where they have to remain for a certain predetermined and designated "time." And then, as if they would have returned to their homeland from a long pilgrimage or a long exile, they experience a foretaste of joy in such a way, which those feel without exception, who were initiated into the Holy Mysteries, mixed with restlessness, admiration, and everyone with his own particular hope.

The demonism of Socrates was the Nous, Spirit or the divine intelligence within him. Plutarch says:

> "The Nous of Socrates was pure and it did not mix any longer with the body, as required by necessity ..."

Every Soul has a certain part from the NOUS, namely reason. Without that, a Human Being would not be a Human Being. There is so much of every Soul mixed with flesh and desire, and when she is changed through worries and pleasures, she becomes irrational. Some Souls immerse deeply into the body, and her whole behaviour is undermined by desire and passion; others are mixed only to a certain part, but the purer part (nous) remains at the same time outside of the body. It is not pulled into the body, but instead, it floats above and touches (overshadows) the outer part of the human head. It is similar to a rope which maintains and guides the remaining parts of the Soul, as long as it demonstrates obedience and is not overcome by the desire of the flesh.

— (Ecclesiastes 12-6)

Figurative Image of how within this World three Worlds in each other,

namely this earthly Sun-World, and also the heavenly and

The outer and the inner Mind
Without God's light you cannot find.

God is free everywhere
Within and without all creatures
GOD
Time measure of Nature
The Angel with six wings
I.

God is the Alpha and Omega
The Beginning and the End
FATHER
Time-Measure of the Law
Lion with six wings
II.

NOON

THE HEAVEN OF HEAVENS

HEAVENLY WORLD CANNOT COMPREHEND

GOD'S Everpresence or essence or Eternity is form Eternity into Eternity

IESUS

GOD'S RIGHT HAND

DAY

DAY

DAY

Point, where Tree of
Life stands.

2 Principium

Entrance to Life.

God's Grace.

GOD
LORD

Division of the Good
from the Bad.

Point, where Tree of
Serpent stands.

1 Principium

Entrance to Death.

DAY

DAY

DAY

LUCIFERS

THE HELLISH WORLD

The way of life is above to the wise,

MORNING

EVENING

NOR ENCLOSE HIM

THE ONE GOD

so that you shun the hell beneath. Prov. 15, 24. cf.

And there is no God
but the one God
H. GHOST
Time of fulfillment
Eagle with six wings
IIII.

MIDNIGHT

Only the Spirit alone knows
Reason in flesh is blind.

God is the first and
the last.
SON
Time of the Evangelium.
Ox with six wings.
III.

the hellish world have their effects. And the darkness cannot conquer the light. It also
shows that the land of the dead, the entrance to hell or superficial darkness, where there
is wailing and gnashing of teeth, as well as the land of the living, the heavenly paradise
or third heaven are from this world. And that the human being has all these things in
his heart; heaven and hell, light and darkness, life and death.

Figürliche Bildung wie in dieser Welt dreyerley Welten in einander, nemlich wie in dieser irdischen Sonnen-Welt auch die himmlische und

höllische Welt ihre Würkungen haben. Und vermag die Finsterniß das Licht nicht. Auch wie das Land der Todten, die Vorhölle oder die äusserste Finsterniß, da Heulen und Zähnklappen ist, sowol als das Land der Lebendigen, das himmlische Paradeiß oder der dritte Himmel, nicht ausser dieser Welt sey. Und daß der Mensch alle Dinge, Himmel und Hölle, Licht und Finsterniß, Leben und Tod, in seinem Herzen habe.

The part that is immersed in the body is called the Soul. But, the indestructible part is called "Nous." The common man believes, it is in him, as he also imagines, that the image that reflects from a glass, is in the glass itself. But, the more intelligent Human Beings know, that it is outside the glass, therefore, it is called a "Daemon" (a god, a spirit).

The Soul flies away swiftly, like a dream, which she does not do immediately after the separation from the body, but later, when she is separated from the intelligence (nous) ... The Soul, which is produced and formed by the intelligence (nous), and even produces and forms the body herself, while she embraces the body from all sides and receives from the nous an impression and form, even though she is separated from the Spirit (nous) and the body, in spite of this, maintains for a long time his figure and resemblance, and due to this, and rightfully so, is called his image.

The Moon is the element of these Souls, because the Souls dissolve in the Moon, as the bodies of the deceased dissolve in the Earth. Those who have been truly honest and virtuous and lead a quiet, philosophical life (without meddling in scandalous affairs) dissolve quickly because they are left alone by the Nous, and they no longer indulge in physical passions. Therefore, they inexorably dissolve and fade away.

In his "Sixth Letter to the Romans," Origenese said:

"There is a threefold division in a Human Being: the Body, or the flesh, the lowest part of our nature upon which the old serpent has buried, through the original sin, the laws of sin, and through which we are tempted to do lustful things; and as many times, as we are conquered by temptation, that is how many times we associated with the devil. The Spirit, in or through which we express the resemblance with divine nature, in which the best of all, the Creator, in accordance with the example of His own Nature, engraved with His finger (that means In His Spirit), the external Law of Righteousness; through this, we will be united (merge) with God and be made "One with God."

As far as the third division is concerned, the Soul has the means between those, who are in a real Republic, to unite only with one part or the other. You either follow this path or that path, you are free to choose, you can choose one side or whatever side you feel most comfortable with. When you relinquish the flesh and you choose the Spirit, then you become spiritual. However, should she succumb to the lusts of the flesh, then she degenerates herself into a body."

The confusion that the translators of the New Testament and ancient philosophical writings have given us regarding the concept of Spirit and Soul, is responsible for many misunderstandings. This is also one of the many reasons why Plotinus and so many other Initiates are now being accused, because they asked for a complete extinction of their Souls – "Absorption into the Divinity" or "Reunion with the World-Soul." Plutarch explained, that only the part of the twofold Psyche, the one which is immersed in the body, is called Soul; meaning of course, the Animal-Soul (to epithymoun). It is the lower, earthly, wanton-principle, in contrast to the reasonable Soul (Psyche).

– *(Compare = Plato, Timaios, Chapt. 31, 32)*

The animalistic Soul (the green Lion) must naturally be dissolved into its Atoms (Prima Materia), before she is capable of fusing her purer essence forever with the immortal Spirit.

The Apostle Paul, subdivided the threefold human nature into Spirit (pneuma), Soul (psyche) and body (soma) and never confused the pure Soul with the principal of wantonness (to epithymoun) – *(Compare Galatians 5-16 and 1.*

Thessalonians 5-23 in the original Greek text).

Paul also taught that there is a psychic body, which is transitory, and a spiritual body which is founded in an immortal substance. The true original text in *Corinthians 15-44* would be, if given quabbalistically and esoterically, as follows:

"It is sown a Soul-Body (not a "natural" body), and it will rise a Spirit-Body."

"Speiretai soma psychikon, eigeiretai soma pneumatikon."

The first Human Being, Adam, was only made into a living Soul (Nepesh); the last Adam however, was made into a live-making Spirit ("egeneto h'o protos anthropos Adam eis psychen zosan; h'o echatos Adam eis pneuma zoopoioun.").   – *(1. Corinth. 15-45)*

Plato remarks, speaking about the Soul (Psyche):

"If she (the Psyche is a Goddess) unites with the Nous (God or divine Substance), then she does everything right and happily; but it is an entirely different circumstance, if she joins Annoia (daemon)."

Even the Zohar teaches that the Soul cannot reach the residence of blissfulness until she has reached the reunification with the substance, out of which she came forth (Spirit). Imprisoned in the body, the Human Being is a Trinity, as long as his defilement is not to such a degree that it does not cause a forcible separation from the Spirit.

Only the lower animal Soul receives from the earthly body such an impression, that even after the separation from the body, she retains its image for a long time. Even when a Human Being is not successful in cleansing this half material Astral-Body of its earthly ingredients for it to join with the Spirit, then after death, this Astral-Body will be separated from the Spirit. Then it is nothing else but an unconscious shadowy image, an apparition, a beingless, senseless phantom, guided only by the impulse of the animalistic life which is extraordinarily strong in him and can last or remain for long periods of time. These phantoms populate the darker sphere of Hades, the so-called Tartaros, where they are facing a gradual dissolution. That is the second death.

Therefore, the second death is inevitable, because the shadowy image which is separated from the Spirit, the animalistic-astral phantom, has lost any memory of its once-divine Genius, and is therefore expunged from of the "Book of the Living." Since it is no longer capable of any repentance or reform, it therefore becomes the object of a total dissolution.

This so-called second death, or the total loss of the Animal-Soul and its Astral-Body, is of course relatively seldom, because in most instances it is possible, even for the departed Souls in Hades, to change their ways and to purify (Purgatory).

It often takes long periods of time until the lower Animal-Soul is dissolved in the Interregnum (in-between Region). By contrast, a purified and advanced Soul that led a philosophical Life, her dissolution is at a much faster pace. Plutarch compared this "Process of Dissolution," the "Shadow of Hades" with putrefaction, and he indicates with this already the occurrence of the alchemistical putrefaction.

This dissolution takes place in the Region of the Moon, as Plutarch writes: Even the ancient Egyptians considered the infernal region to be on the dark side of the moon. Whatever the moon represents in the solar system, is the fantasy or the power of the imagination of a Human Being.

His earthly and animalistic concepts and wishes are born out of his fantasy, which disappear through the Light of Recognition, like fog before the sun.

The earthly Human Being is the Son of the "Moon." Even when his external sensory life appears to him as much as it can, as a reality, it is nevertheless nothing else but a confused dream-life and his self-consciousness is nothing but a delusion. Not until the earthly Human Being (the old Adam) has ascended into a divine Human Being (Christ), does the Son of the "SUN" (that means = God's) come into the consciousness of immortality, and the true life begins for him.

This Re-Birth out of Water and Spirit is only possible, if and when the "transitory and earthly" in a Human Being are dissolved beforehand. That is why, the main degree of Alchemy is the dissolution or putrefaction. The animal Soul fears and rejects nothing more than this dissolution, and for her, it can be compared to actual dying. That is why these earthbound Souls cling so much to the external material life, and they cannot, even after death, separate from their earthly body. Instead they roam around the area in which their former companion of their "Pleasures" lies buried.

Plato describes the difference of the destiny of a pure Soul and an impure Soul, after death, as follows:

"And the Soul, the Invisible, is drawn to a place which is similar to her nature, a noble, pure and invisible place, the true Hades, to the good and wise God. This Soul, which is created and disposed of in this manner should, after the separation from the body, immediately disperse and perish, as most Human Beings assert? Far from it. On the contrary, it is more like this: When the Soul separates from the body in a clear and pure manner, without taking anything with her from the body, and if the Soul had nothing voluntarily in common with the body during its lifetime, but instead she fled and drew back into herself and was always mindful in regards to that, that of course, has no other meaning than, that she properly philosophized and prepared in truth for an easy death. If she is in such a condition, then she will go to the invisible, to the divine, to the immortal and the sensible, which is similar to her. When she reaches this point, blissfulness will be bestowed upon her, and she is liberated from error, ignorance, fear and all the wild passions of love and all the other human evils, while in reality, as we are being told by the initiates, she lives united with the gods for the remainder of time.

Should the Soul however, separate from the body stained and impure, because she (the Soul) was constantly together with the body and cared for him (the physical body), and loved him and was fascinated by him and his wantonness and desires, and nothing else appeared to be the truth or reality for her other than the material, the physical, that which could be touched, seen, eaten, drunk and used for the satisfaction of love, and if it was her habit to hate, avoid and fear what for the eyes was enveloped in darkness and invisible, and what on the other hand, was according to reason and comprehensible through Philosophy; when a Soul is in that condition, she will not be able to separate in a clear manner."

If instead the Soul is penetrated with the corporeal, with which she virtually grew together through the association and the dealings with one body, and because of being together uninterruptedly with the body and also the great amount of attention which was paid to the body.

This corporeal (physical) is something depressing and burdensome, something earthly and visible, so that even the Soul, which is burdened with this, feels its weight and is drawn back again into the region of the visible, out of fear for the invisible and Hades. Here she roams around the monuments and graves. In this vicinity, many dark apparitions of Souls have been seen. They are Souls which did not separate in total purity from the body, but still take part in the visible. That is the reason why they can still be seen."   — (Plato, Phaidon)

That is why it is very important that the Human Being, even

before his death, pays attention to the life in the other world; to the life which is yet to come. Plato explains this when he writes:

"If the Soul is immortal, then she requires loving care, not only for the present time which we call Life, but also for the entire time, and that is why the danger appears to be very great when someone neglects his Soul. If death would be a separation from everything, then it would be a gain for the evil ones who, when they die, dispose their body and would, at the same time, dispose of the wickedness of their Soul.

The Soul has proven itself to be immortal, and there is no other protection from evil and no other deliverance than the endeavour to be as good and reasonable as is possible.

The Soul cannot have anything with her when she comes to Hades other than her education and discipline (rearing), which will bring the deceased the greatest benefit or harm, immediately, in the beginning of the journey there. It is said that the demon of every deceased Human Being, which during his life on earth had the Human Being under his protection, will seek to lead the deceased Human Being to a place from where the ones who are gathered there, after they have been judged, will travel with that guide into Hades, who has orders to lead the deceased from here to that particular place. After they have received what is due to them, and they have remained there for an appropriate length of time, another guide brings them back here again, for many and long periods of time. That Soul, however, which is desirously attached to the body, flutters around the body and the visible area surrounding it for a long time, and only after considerable resistance, much suffering and with much effort, will she be led away

157

forcibly by the demon, who is ordered to do so. The good-mannered and sensible Soul follows willingly, and whatever happens to her, is not foreign to her."

— (Plato, Phaidon)

All this will only be understood in the proper manner, if you are truly cognizant of human nature with her metaphysical tendencies. The present-day "Pastors" and "Psychologists" do not know, in reality, what the Soul actually is, because they do not know the true esoteric Psycho-analysis, which was already known to the ancient Egyptians.

Without these teachings of the so-called "Seven Souls" of the Egyptians, the ancient Egyptian Sarcophagus and Pyramid texts cannot be understood, nor can their mysterious Book of the Dead. The Greek, quabbalistic, hermetic and original Christian Mysteries will also remain incomprehensible. In order to give the true seeker the key to these mysteries, the following statements should be studied in particular detail.

The Egyptian Initiates taught that the Soul, after the death of the physical body, had to travel through seven chambers or principles, because they divided human nature into seven spiritual-psychic potencies, as did the Greek Initiates and later the Quabbalists. The seven basic parts of the Human Being are:

	Greek	Egyptian		Quabbalistic
1.	Divine Nature, divine Spirit, the Father, the Light. Zeus	*Chu* Amun-Ra	♃	Jechidad
2.	The divine-spiritual Nature, the Nous, divine Intelligence, Wisdom, SON, Christos-Dionysos.	*Chaib* Horus	☉	Chaijah
3.	The intellectual Nature, Reason, Logos, Hermes.	*Ba* Thot, Anubis	☿	Neschamah
4.	The sensitive, perceptual Nature, the human Soul, Psyche, Persephone.	*Ab* Osiris	☽	Ruach
5.	The instinctive, passionate Nature, Anoiathymos, Animal-Soul, the old Adam, Kain, Eros.	*Ka* Set-Typhon	♂	Nephesch
6.	The half-corporeal astral Nature, Evestrum, Mumia, Astral-Body, Image, Eidolon, Eva, Demeter.	*Anch* Hathor Isis	♀	Habal-Garmin
7.	The material, physical Nature, earthly Elementar-Body, Pluto etc	*Chat* Geb	♄	Guph

Only on the basis of the above chart, can you understand the sense of the ancient Egyptian text, where it is said, that you should join the Ka with the Ba, that means, the instinctive Nature or the animal Soul with the Reason, just as the Greek Mysteries taught.

Just as the Book of the Dead reports a succession of transformations, which the departed Soul has to go through until the reunification of Ba and Mumia (Mumia, Evestrum) and this takes place under the direction of Anubis.

As a matter of principle, the following must be mentioned:

Spiritual-psychic and physical purification are, for the Egyptian, originally one. A new Form of Existence must be inserted psychically, as well as physically. This transformation is the true mysterium.

The Body (Astral-Body) is being prepared by Anubis (Hermes). His symbol is the animal-hide, which he bursts out of at the time of death. Ba, at the same time, experiences a succession of transformations. He passes through all kingdoms (Earth, Water, Air) to eventually, when his substrate the Mumia is restored again, he will again unite with it.

"He, the Ba, looks at his Mummy, upon which he reposes, without ever perishing."

— (*The Book of the Dead, Chapt. 154*)

The resurrection must therefore, also take place in the body (Astral-Body), which also has to be renewed and "restored."

The word "Mummy" should, under no circumstances, be mistaken for the external visible corpse, but for the vegetative and vital principle "Anch" (the vital energy, the "Sycamore of Hathor," that means, the Tree of Life) which is hidden therein. That is why it is written in the ancient Egyptian texts,

"Osiris died in the Nile, in order to rise in the plants to a new Life."

In the Egyptian hieroglyphics, there is a great amount of representation where, out of the Osiris-Mummy, plants grow forth.

The union of Osiris (Soul) and Isis (Astral-Body), through which Horus is being procreated with the new Sun or Spirit-Body. Osiris is the Water (the Moon), and Isis (like Demeter), the

160

Fertility of the Earth. That is why she (like Demeter), is considered to be the Donor of Wheat. Therefore, the wheat kernel is an ancient egyptian symbol for the resurrected body.

A Verse about becoming a grain:

"I enter as a grain and come forth as bread (Body). The gods live within me, I live in the grain, I grow in the grain, which the Heavenly sow, hidden in Geb" (that means, in the Earth, in the earthly body). – *(Sarcophagus Text Nr. 58)*

"And that which thou sowest, thou sowest not that body, that shall be, but bare grain ..." – *(1. Corinth. 15-37)*

The Egyptians did not only undertake the function of embalming of corpses, but this also took place at the cult of the gods, during the coronation or initiation. The ointment represented the female principle; that means, the Isis, which tied together the members or limbs of Osiris, the alchemistical Gluten (bonding agent).

The Mummy (from the Pers. mumija) is a wax, a soft, balsamic resin or gum, a very precious, fragrant, mountain-balsam for the healing of wounds which the Egyptians used for mummifica'tion. The corpse was embalmed with this Mumia (Ointment), that means, totally saturated, and thusly the corpse itself became Mumia.

"Hail thee, my Father Osiris, I have come to embalm you ... You embalm my body. I am totally immortal like my Father Amun. You let me descend to the Land of Eternity, as it happened to you, together with your Father Atum, whose divine body will not perish, nor will it dissolve."

– *(The Book of the Dead, Chapt. 154)*

For the Egyptian, the Mumia is never a corpse from which the Soul has departed, but where, to a certain degree, the Soul is present. Because, as long as the earthly body has not decomposed, the aetheric Astral-Body (Habal-Gamin) remains to be joined through a magnetic cord, through which the body obtains certain magical attributes. The earthly body is, to a certain degree, saturated through mummification with a remaining emission of the Astral-Body. This is also what Paracelsus understood in regards to the Mumia; this aetheric-astral "Balsam of Life," which he attributed with certain magic energies. About this, he very carefully expressed himself:

"Great Secrets are within this, which however, better remain concealed, because a great amount of mischief can be caused with this."

— (Paracelsus, Tractate of Philosophy III, IX)

The Egyptians were also convinced that the aetheric-astral principle (the Astral-Body) remained in magnetic contact with the mummified earthly body. In this connection, it should not be overlooked, that only the bodies of high initiates (Pharaohs and High Priests) were embalmed and not the common Egyptian. It is, without a doubt, that the Egyptian Priests wanted to maintain through this, the high-white-magic emissions, which emanated from the mummies of the great Initiates for long periods of time, to enliven and strengthen their religious Mysteries. The mummified bodies of the Initiates were, at the same time, the accumulators of this magic energy, which flowed through the Egyptian Temples. There were also reports of such similar mumial miracles from the departed biblical Saints. The following is said about the Prophet Elisha:

"And when the man was led down and touched the bones of Elisha, he revived, and stood up on his feet" (that means, a dead man by touching Elisha's body, became alive). — (2. Kings 13-21)

The same is written in the Book of Sirach:

"He did wonders in his life and at his death were his works marvelous." — (Sirach 48-14 or 15)

At this point, we should be reminded of the resurrection of many of the Saints, which are reported in the Gospel of Matthew, Chapt. 27, Verse 52-53.

We should be aware of the fact, that the physical corpse living in the grave was not meant with this, but only the Habal-Garmin which was connected with it, that should be totally clear a priori. Only the maintenance of the latter was of importance to the Egyptian initiates. That was the reason for the many festive Sarcophagus and Embalming Rituals, which in reality, were nothing else but Initiation Rituals. The Embalming of the corpse was only a repetitive Initiation Ritual: There comes Horus with the salve (ointment) and embraces his Father Osiris, when he found him prostrated. And Osiris was fulfilled from that, what his only begotten Son did for him ...

> "Oh NN, I come to you – remain! Be fulfilled from the salve, which comes out of the Horuseye (Light)! Be totally fulfilled by it!
>
> It binds your bones together and unites your members; it holds your bodies together and eliminates your bad secretions ... Acquire its wonderful fragrance, so that you are sweet-smelling, like Re, when he comes forth out of the Light-Sphere. The gods of the Light-Sphere are happy about that."　　　　　– *(Pyramid Verse 637)*

The point is, if Osiris (the Soul) unites with the Balsam of Life of the sweet-smelling Isis, that means with the cleansed Astral-Body.

> "Because of that, put our red servant (Gabritius) to his sweet-smelling sister (Beja)."　　　　– *(Radix Chymiae)*

> "OH Osiris, with you together comes your Ka, and blessed are you in your Name (Ka-hotep). He illuminates you in your name of an "Enlightened One."
> 　　　　　– *(The Book of the Dead, Chapt. 128)*

> "Wake up! Oh NN, rise up.... shake the dust from your

limbs.... receive the white shining linen cloth, your Light-Garment." — *(Pyramid Verse 414)*

It is the Garment that, as Plutarch says, has nothing shadowy, nor coloured, but it is Light:

"Because the beginning is unmixed and the original thought is pure." — *(Plutarch, De Iside et Osiride)*

The following addendum is proof that Plutarch is describing at the same time the attire of the Isis and Osiris Mystics:

"The reason why the Osiris Garment is only worn once (namely only during the highest Initiation, the Epoptie), it is then taken off and from then on, it is kept unseen and untouched. Whereas, the Isis Garments are worn very often (namely for every festive occasion). These sacred garments exalted the wearer at the moment of the highest initiation, almost to the level of a god, and also guaranteed his resurrection from the Mortal-Earthly to the Immortal Divine. These holy garments were kept by the Initiates as a shroud (Garments worn by a dead person). The Mystic who is buried wearing these sacred and divine garments could be assured of transfiguration and resurrection. Those who were especially Chosen (God's Elect), the transfiguration would become visible on the now sanctified corpse. At the time of his death, the great Theurgist and New Platonist Heraiskos was wrapped in the holy Osiris Bandages. According to the Damaskios report, upon the Linen Bandages were brilliantly shining mysterious Symbols, and all around them divine appearances were seen! Heraiskos had now himself become "One" with the killed and resurrected God. This Osiris Garment belonged also to the twelve mystical garments, which the Initiate would wear once he had reached the highest degree of initiation in his life, to which the Egyptian writings explicitly refer.

You would think it was a contradiction when the Egyptians taught that the Soul should be purified and liberated of everything which is earthly, and on the other hand were trying to artificially retain the Soul by mummifying the corpse. That is not so. As previously mentioned, only those who were already Initiates were embalmed – those who were already born-again before their death. Through the Re-Birth, their bodies (their Astral Body) had become holy and transfigured, so that they contained nothing earthly any longer and they were no longer bound to their coarse physical or material body. They could go in and out of their material body as they pleased, as it is written in the "Book about the Going-Out of the Day":

"To let him ascend and descend according to his wishes, and that he can do anything unimpeded, whatever his senses desire." — *(The Book of the Dead, Chapt. 163)*

The impurities do not lie in the corporeal (in the Astral Body), but in the Animal Soul (Ka). That is why the purification of the Animal Soul is referred to again and again:

"Your Purification is the Purification of the gods, which are gone to their Ka. You are pure, your Ka is pure, you Ba is pure … Horus cleanses you with the heavenly moisture (the Life-Water)." — *(Egyptian Pyramid Texts – Verse 452)*

"The Light of the body is the eye. When this light is pure, your whole body shall be full of light." — *(Matthew 6-22)*

"The Earth is in its own nature, a beautiful crystalline body, as light as the Heaven…. Consequently, it is clear, so that the darkness comes to the fire."
— *(Thomas Vaughan, Aula Lucis)*

The Mystery of Alchemy.

The Central-Secret of Alchemy is the threefold Re-birth of the Spirit, the Soul and the Body. The mysterious occurrence stands in a certain parallel to the ancient Mysteries, which were discussed in great detail for this particular reason.

The Lapis Philosophorum is the Mystery of the inner Salt-Body. The whole Magnum Opus of Hermetics revolves around the dissolution, transformation and spiritualization of the same. It is the impure half material Soul-Body (soma psychicon), who through a succession of very secret processes, should be changed into pure Spirit-Body (soma pneumaticon).

The hermetic sciences have, in order to describe these mystical processes, adopted many pictures out of the myths of antiquity, and have taken out of the domain of chemistry a number of symbols. Thusly, a peculiar compositum of technical expressions came into being, that to the uninitiated gave cause to the greatest variations and to the most absurd or nonsensical interpretations.

However, the Initiates knew that as far as Alchemy was concerned, they were dealing solely with the Mystery of the threefold Re-birth of Water, Blood and Spirit, whose secret processes the disciple of the hermetic arts had to pass through by himself. Just as once the "Mystic of Antiquity" experienced in a mysterious manner on himself the destiny of dying and being the resurrected Son of God.

This Re-birth and Resurrection takes place in the living Human Being and not in a dead corpse. Because this incorruptible resurrected body is born of the immortal Spirit, and not from corruptible matter.

It is a great error to believe that after death this spiritual body, in some miraculous manner, comes into being by itself. Just as it would be in error to assume, that this Spirit-Body would already be in existence in us. The only thing left to do would be to slip out of the coarse material earthly body in order to continue to live in the World Beyond in the more subtle immortal body.

What is, however, already in existence in us, is the so-called "Soul-Body (soma psychicon) of the Apostle Paul. That means, the impure half-material Astral-Body, the Mumia of Paracelsus, the Habal-Garmin (Bone-Spirit) of the Quabbalists which, however, contains too many impurities from the earthly matter, and that is why the Soul-Body in the "World Beyond" is inevitably subject to dissolution if he is not purified during our lifetime and transformed into a transfigured Spirit-Body.

The half-material Astral-Body is the Doppelgänger of the elementary physical Body. He builds and nourishes himself of its finest substance, the same as the plant nourishes and builds itself up out of the finest energies of the Earth. The Astral-Body, which is being placed between the Animal-Soul and the Earthly-Body does not only take its nourishment out of the substance of the latter, but takes even considerably more out of the substance of the lower Animal-Soul, since the half-material Astral Body is the connecting link between the lower Animal-Soul and the earthly body. That is why he is actually the image (eidolon) of the Animal-Soul, which formed him, while he owes the earthly body only to a part of his half-material substance.

Here on Earth in the coarse earthly body, we Human Beings can hide from each other as if behind a mask. Through death this outer shell falls away, and everyone's true form or shape (the naked form of his Soul-Body) shows, and nothing can be covered up or concealed any more.

Whosoever wants to have a pure and well-built form in the life beyond, should make the effort now in this life, to develop a noble character. All psychic impurities must be cast off and in our innermost we have to let the Divine emerge, so that the Soul-Body can become transfigured and immortal. Then the God-

Man can be born in the Soul and can take shape and form, which no longer represents the image of the animal, but the re-established image of the heavenly. — *(1. Corinth. 15-49)*

Because, when the Human Being becomes again the pure Image of the **Elohim**, which was lost through the lower Adam, then he becomes a totally new creation in **Christ**.

In the true sense of the word, it should be the task of every Human Being to be an "Alchemist," to set into motion the potentially higher endowed psychic-spiritual energies, which are in every Human Being. To transmute his lower nature (that means his lower Animal-Soul inclusive of the Astral-Body) into a higher spiritual, and to produce in this manner within himself an immortal Spirit-Body. Then out of the "lower metals" the imperishable (immortal) "Gold" of Sages is made, and the corruptible is transmuted into the incorruptible. — *(1. Corinth. 15-53)*

What is of importance here is solely the hermetic practice, the alchemistical process.

Even when the Alchemists openly elaborated in their works with regards to the different symbolic designations and their three principles (Mercurius, Sulphur, Sal), and it was clearly explained to a certain degree, it does not apply when it comes to the process of the Magnum Opus. This mysterious process is being explained theoretically in the writings of the Adepts in the ancient alchemistical literary language, but nowhere is it clearly explained when it comes to the true practical significance.

The ancient Masters of the Sages did not write for the great majority, but only for the few true students of the hermetic sciences, and friends of the hidden wisdom.

They wrote about the Mysteries and as it is when it comes to all Secret Orders, leagues or covenants and associations. They were bound since time immemorial to recognizable signs in regards to certain names, designations and expressions, to which only the initiates alone could associate the actual and true meaning or sense. Even the initiated members themselves would change their names (as it occurred when someone became a monk) and upon admittance into the high Order, they received a new name. In this manner, the ancient Sages wrote as Epopts and only for the Initiated, and separated through their unusual terminology, their secret philosophy from the common.

When the Archon Nicodemus went, in the night, to Jehoshua, the Prophet of the Nazarenes, he was taught by Jehoshua about the necessity of Re-birth out of Water and Spirit. Nicodemus asked the serious question: "How can that occur or how can these things be?" That means, how is this process being carried out. Upon which Jehoshua gave him the following answer: "You are a Master-Teacher (Rabbi) and you do not know these things?"

In the Gospel of John, the mysterious proceedings of Re-birth are not explained in detail, but are only indicated to Nicodemus, that no man will ascend to heaven, as only the one who descended from heaven to earth, namely the Son of man, who originates from Heaven. This Son of Man, Jehoshua continued to teach him, He must be exalted again by the Earth, the same as Moses exalted a serpent in the desert, which as the Book of Wisdom says, is identical with the Son of Man.

– *(The Wisdom of Solomon 16-7)*

In answer to the question, who this mysterious "Son of Man" is, Rabbi Jehoshua replied at a different part of the Bible:

"Yet for a little while is the LIGHT in you. Walk as if you have already obtained the LIGHT, so that the darkness will not overcome you. Whosoever walks in darkness, cannot see whither he goes. Believe in the LIGHT, while you have it in you, so that you become "Sons of the Light."

– *(John 12; verse 35-36)*

The great First HUMAN BEING (Adam-Kadmon) is GOD, the FATHER Himself, whose very Image or SON is the pure God-Man (Christ).

"And above the heaven that was over their heads, was a likeness of a throne, as the appearance of a sapphire stone: and upon the likeness of the throne was the likeness as the appearance of a man above upon it." — *(Ezekiel 1-26)*

"I saw in the night visions, and behold, one like the Son of Man came with the clouds of heaven, and came to the Ancients of days, and they brought him near before him." — *(Daniel 7-13)*

"And in the midst of seven candlesticks one like unto the Son of Man." — *(Revelation of St. John 1-13)*

In the Human Being itself the "SON" of the mysterious **Archetype Human Being** is being born.

"The great secret of the Samothracians, which is unutterable and which only the Initiates know. They however have knowledge in the greatest of detail about ADAM as their First Human Being." — *(Church-Father Hippolytos)*

Before the fall, the heavenly Adam was the pure image of the divine Father, the Proto-Adam. After the fall, this image (Zelem Aelohim) had become darkened in Adam, and he himself became "Adam Belial" (fallen Adam). That is therefore the one in us, the Son of Man, who descended from heaven, who must again be exalted to his first dignity and then he should ascend to heaven.

Moses (the divine Genius) exalted a serpent in the (inner) desert. It was the same serpent, through whose suggestion Adam turned away from **God** and fell; namely the serpent of reason and the intellectual nature (Neshamah). Originally, this serpent was heavenly and had wings. But, he fell first, and through this, he lost his wings and became an Earth-Serpent.

"I beheld Satan as lightning falling from heaven." — *(Luke 10-18)*

"And the great dragon was cast out, that old serpent called the Devil, and Satan, which deceiveth the whole world; he was cast out into the earth." *– (Revelation 12-9)*

Because, the meaning of "nachath" in hebrew is "descend," "neshath" on the other hand means "to become exalted," "ascend."

The serpent (Nachash) is therefore the descending Spirit, which nourishes itself from the dust (Aphar), that means, Earth-Spirit. Neshamah, on the other hand, is the ascending or heavenly Spirit, the exalted serpent!

Adam himself however, should be dissolved to Adamah (red Earth) again, from the dust (Aphar) from which he was made.

In order to properly understand this, one must know that there are two Adams: the archetype heavenly **Adam (Christ)** as spiritual image of **God** (Zelem Aelohim), and the "Adam made out of dust," the earthly embodiment of the first. Only the latter fell, and his fall was caused by the loss of his divine image (Zelem Aelohim).

Adam means "Blood." The body's life (Nephesh) is in his blood. *– (Leviticus XVII-14)*

Adam is therefore the in-the-blood living and moving Animal-Soul (Nephesh). In hebrew, Adamah also means "red Earth," but the external coarse Earth (Arez) is not meant by that. Instead the subtle spiritual Earth, a red tincturial Gold-Dust, a sulfuric-fixed Spirit-Salt, whereof **God** created the etheric body of Adam in paradise. This Aphar min ha Adamah is the same substance out of which, in the inner of the earth, Gold comes into being. That is why it is also called the primum ens auri. This sulfuric emanation of the Sun, this fleeting Gold of the Philosophers, the incombustible red Nature-Sulphur is intimately united with the incorruptible Nature-Salt, and becomes fixed in the form of a shining kernel or glittering body (corpus solare et lunare).

Therefore, Gold is nothing else but a glittering, yes rather red clay or yellow clay: Adamah, being made to glitter through the Nature-Salt. *– (Limus Terrae Catholicae. Khunrath de Chao, p. 194)*

Take now, or separate from this (through that in Mercurium Philosophorum resolved moist Nature-Salt this lustre; which in the beginning was overcome through a now-conceived Salt), and that is how you get red (incombustible) beautiful mud, or (provided this mud would be at the same time with the Nature-Salt coagulando figiret) what would remain would be an exquisite (highly red) Tincture-Powder and a tingible Gold-Stone.

– (Fr. Beissler, Tree of Life, 1683)

Adam or Adamah is also a subtile highly-red solar earth, a coagulated Sun-Fire, a red sulphuric smoke (Ruach), the Calidum innatum. Aphar on the other hand is a subtile lunar earth, a coagulated Mercurial-Water and fixed Silver-Salt, also called Root-Moisture.

Rabbi Hunna said:

"When Aphar (= dust) *(Genesis 2-7)*, is mentioned, then it is female earth; when Adamah is mentioned, then it is male earth. The potter puts together male and female earth, so that his vessels become strong."

– (Midrash Raboth B'reshith Vol. 12)

"That is why we say, that in this World there are four lights, two heavenly and two central; the heavenly, Sun and Moon, everybody can see, the central lights however, are covered – one with the earth, the other with the water. That is why it is not believed that they exist. That the fire concealed in the earth is of a solar nature, a little coarser than the Sun.

In the water however, is a thick air of a lunar character, but not as light as the Moon. The central Sun throws a heated male Salt (Adamah) into the water; the water accepts it and adds to it its female mucous semen (Aphar), both of which rise with the air. That is how the semen of the body is prepared. The semen of the body also receives from heaven its life, from the Moon the Spirit, and from the Sun the Soul. In this manner the four lights come together, since the heavenly give Spirit and Life, and the earthly give the body. This semen is carried by the wind (Air) in a concealed manner and revealed in the water, namely in the light crystalline water, out of which it is drawn, because there is nothing else under the Sun wherein it could be found." – *(Eug. Philaletha, Euphrates)*

Through the fall of the Human Being, this adamic earth is being cursed and darkened. First the divine Spirit withdrew from the Human Being, and his Spirit lost the intimate bond with God. At the same time, his Soul (Ruach) withdrew from the Spirit (Neshamah) to the outside, and the Nephesh outside of the Ruach. With this, the inner firm connection was loosened between the Neshamah (Spirit), the Ruach (Soul) and the Nephesh (animalistic Principle), and Adam's etheric Light-Body transmuted partly into an external material Body.

Should this curse be terminated and the whole instance be rescinded, then the etheric Body of the Human Being, which became half material and enclosed in him, darkened body-Soul (Nephesh) must be dissolved again into its prima materia, that means, the Mercurial-Water.

The well out of which the Sages draw this water is very secret and it is very much concealed to the uninitiated.

Alchemist: "You only speak in riddles – tell me, are you the Well, of which Count Bernhard of Treviso writes?"

Mercurius: "I am not the Well, however, I am the water. The Well encompasses me everywhere."

— (Svendivogius, Novum Lumen chymicum)

Our Gold-Body is of no value to us, if he is not totally dissolved and brought back to his beginnings; that means the Chaos of the Philosophers. All Gold was once Silver. Originally, the psychic Body was Spirit, and therefore he must be dissolved again into Spirit, if he should rise in a new transfigured form.

Also, the instinctive Nature-Soul (Nephesh) enclosed in him requires dissolution and rebirth. The dissolution of the body occurs, as it does with the Soul in their own sphere, namely, as Plutarch mentioned, in the astral region between the Moon and the Earth.

According to the Quabbalah, this region is dark and desolate, that is why the departed Souls (according to their condition) have to pass through the dark path for three days. At the end of this path, the path divides into two paths, of which one leads to the upper Gan Eden (Paradise), and the other path leads to the lower G'hinam (Gehenna), that means, into the Tartaros.

At the dividing point of both paths, you will find many kinds of dark demons, and at the top the "snarling Dog" (Cerberus infernalis triceps, the Guardian of the Threshold), the one who will immediately seize a Soul, once he sees a Soul that belongs to him.

This dark region of the interregnum, which is inhabited by countless roaming Spirit-Beings, represents a great shock for the Souls travelling through this region. However, the pious are protected through their pure holy garments and through their psychic Signature that they wear, so that no dark Being can be in their proximity or come close to them.

For this reason, it should be obvious why the dissolution, purification, and rebirth of the psychic body for the Human Being, who has gained this understanding, is absolutely necessary. In the Mysteries of Antiquity, those who were to be initiated were artificially transferred into a state which would be identical to what the common Human Being will experience after death. In this manner, it was possible for them, on the basis of this profound mystical experience, to transform and totally restructure their inner life, so that the Initiates look forward to the end of their earthly existence with philosophical clarity and cheerful calmness.

The Mysteries of Antiquity no longer exist, but in its place are now the Mysteries of Alchemy, which teaches the great Arcanum in the Picture-Language of hermetic Symbolism.

It is the Arcanum of the Rebirth out of Water and Spirit, and closely connected with this is the ancient, but publicly never-answered Nicodemus Question: "How can these things be?" "Can a Human Being enter a second time into his mother's womb and be born?"

A Human Being cannot only, but must, to be born again, be led back into his prima materia, out of which he came forth according to his astral nature.

Before the "Son of Man" can be exalted, he must be humbled, and before he can ascend to heaven, he must walk for three days through hell. *– (Matthew 12-40).*

This is the condition of the alchemistical Blackness, whose Central-Darkness, as Ripley taught, extends over three days and nights, even though the total blackness last for 40 days.

Even Jesus remained for 40 days, after His resurrection, on Earth before He ascended to Heaven. *– (Acts 1-3)*

"Therefore, you must go through the Gate of Darkness if you want to win the Light of Paradise in the White."

– (Ripley)

Then the psychic body has lost its old form and takes on another. This form is the ash, that means the Kings and the Queens (Sol and Luna) grave. Whosoever destroys the Gold in such a manner, sowest a good seed into good earth, out of which he comes forth again with a hundredfold multiplication. But whosoever wants to keep his Gold, looses his effort and labour and will be cheated.

The descriptions of the Philosophers who dedicate this condition to the blackness are mostly allegoric.

This is what is being said with respect to this:

"Take two serpents, which you can find anywhere on the ground, one living male and one living female. Join both, with the bond of love, and lock them up in the arabic Caraha. This is the first work. With the Fire of Nature you must struggle against them and you must take care, that you draw your line around them. Encircle them, and keep all entrances secure, so that they cannot obtain any help. Keep it under siege and be patient, then they will transmute into an ugly filthy poisonous black toad, which will change into a horrible guzzling dragon, who will creep and wallow on the bottom of his cave, but without wings. Do not touch it, because there is no poison, which is stronger or more potent. Continue as you did in the beginning, then the dragon becomes a swan, whiter than snow. Then I will allow you, to increase your fire, until the Phoenix appears. This is a dark red bird with a shiny fiery colour. Nourish this bird, with the fire of his father and the ether of his mother; one is his food, the other his drink. Without the latter, he will not reach his total magnificence because the fire does not nourish well, when it has not been nourished first. In itself, it is choleric and dry: but it is being tempered by a comfortable moisture, it provides it with a heavenly complexion and brings it to the required exaltation. Therefore, nourish your Phoenix, then he will move in his nest and rise like a star. Therefore bring Nature into the horizon of Eternity."

– (Eug. Philaletha, Lumen de Lumine)

The principal matter however is this: to make the stone into water, that means, into aqua physica pulverulenta or prima materia. This is such an extraordinary important preparatory work, not much is being said about this in hermetic writings. Only sporadically can you find scattered information such as the following:

"In the preparatory work, you must separate Body, Soul and Spirit from each other, purify and reunite them. Be only concerned about our water and the foliated earth, the Spirit cannot be seen; he moves all the time above the water. The foliated earth is like a little island in a philosophical ocean. You must crush and close up this earth, then it will, because of being thirsty in prison, shatter by itself and it will, like a thick water, mix with oil. You must know how you unite this, as terra foliata, in proper proportion with the water. Pondus aquae esto plurale, terrae vero foliatae singulare."

— (Hautnorton, *Tractate of the philosophical Salt, 1656*)

There is so much uncertainty, even in the circles of the hermetic researchers, with respect to the threefold order of the Stone of the Philosophers and the true significance of the projection.

The highly concentrated Lapis is, as it has been explained many times, the magic-active Fluid-Body. The Stone of the first Order tinges spiritually, that means only spiritual projection can take place. It is known, that certain Magicians and Rosicrucians, for instance the Alchemist Prestel, was capable of transferring the pictures of his imagination consciously to other Human Beings, so that in that manner, they could see these things which were not present in the outside World. There are reports that Gautama Buddha had also these same abilities.

The Stone of the first Order makes the so-called spiritual or Spirit Alchemy possible.

The Stone of the second Order tinges spiritually and psychically. With it you can affect psychic projections, for example, the casting out of devils, healing of psychological ailments etc. The Lapis of the second Order has the form of an oil or Elixir respectively, Aurum potible, and makes a psychic or Soul-Alchemy possible. To this order belong Jesus Christ's and His Apostles' Miracles of Healing.

Finally, the Stone of the third Order tinges physically. Therefore, material projection can be done, such as transmutations of the elements, refinement of the metals, transmutation of substances, healing of physical ailments, rejuvenation and transmutation of the body, up to transfiguration. The Stone of the third Order has the form of a powder (Aphar), with which the projection of the physical Alchemy can be realized. Only fully developed Alchemists and Rosicrucians are in possession of the Lapis of the third Order, which lends magical power even over external Nature.

"That is how in a Human Being, that is perfection, of what we dealt with, yes and much more, because his eternal living Spirit is immediately a spark of the living Divinity. Therefore, a Human Being should learn to recognize himself, then he will be able to judge everything out of his nature; even the four separated principia or qualities secundariae, on the whole they are called the Elements, all the creatures which are invisible to us, will be on the whole, totally discovered and revealed; yes Heaven and Earth, Light and Darkness, he will totally restore even the most dead corpus and make it eternally permanent. Yes, he will recognize, how once all earthly creatures will return again from the darkness and be transferred into a spiritual mercurial Life. Therefore, oh Human Being, learn to know thyself well, then nothing will be hidden from you, whatever it may be, the Philosopher's Mercurius, Moses but

Fire, that is, what is called Salt and Water, you will understand, what is written in Jeremiah 10-13: "He maketh lightning into rain." And because we are ourselves what we seek, it is therefore only righteous, that we make the beginning in us and on ourselves, to reach the recognition of the original state of the whole creation and for the recognition of the Creator Himself, to Whom we give Praise and Glory in Eternity." – (G. v. Welling, Opus mago-cabbalisticum)

In conclusion, have the knowledge that all gifts of perfection come from the Father of the Light, and that Wisdom cannot enter into a depraved Soul, despite reason and erudication! The Highest is inclined to those with mercy, those who fear Him, those who love and call upon Him with a sincere Spirit:

"Without the true preparation, nobody will even reach what is considered to be Alchemy."
– (From the Tractate "The Light Which Breaks Through The Darkness," 1772)

Catalogue of Publications

The publications of Merkur Publishing Company Limited are specialized, but not limited to the fields of Philosophy, Medieval Studies, Medieval Medical writings, Metaphysics, Christian Studies, Hermetics, Classical Herbology, Classic Alchemy, and related studies.

We have endeavoured to keep our translations as accurate as possible. Diligently translated from old German, Latin, and Greek some very very old, others are more recent. These works follow very closely the Laws of Nature as written by such scholars as Paracelsus, Plato, Socrates, Roger Bacon, Robert Fludd, Hermes, Nostradamus, Lao Tse, Karl von Eckartshausen, and the Bible.

Our purpose is to bring attention to old and significant works of written Wisdom for the benefit of all.

Philosophia Mystica
    The Prophecies of Daniel
    Author: Paracelsus
    ISBN 0-9693820-0-6
    Date: 1989
    Format: 158 pages paperback
This work was written originally in the Swiss-German dialect of the 1500s and translated from the original text. Among all of Paracelsus's writings it was one of the lesser known, but not of lesser importance. Paracelsus explains, through the principle of cause, effect and consequence, the troublesome problem all Human Beings face in regards to their negative character-traits (= actual Sin) and how to dispose of them (= true repentance) thereby leaving bondage and entering into freedom (= true forgiveness). The process described by Paracelsus in this book will lead the reader to purification and thereby making the first step into Hermetics.

The reaosns for all of our daily worldly concerns and problems and the confusion they create are explained. However, Paracelsus is always very blunt and candid in his approach, his style, requires attentiveness on the part of the reader. Wisdom being simple, it is often illusive and appears to many as paradox.

The latter part of the book explains the Prophesy of Daniel, from the Holy Scriptures. The accuracy of the Prophesies of the Prophet Daniel is astounding, even as to present day events (Persia, Yugoslavia, Russia, Germany). As to the future: Wales?

The Principles of Higher Knowledge
Author: Karl von Eckartshausen
Date: 1991
ISBN 0-9693820-1-4
Format: 311 pages paperback

Originally published in the late 1700s. This book is a treasure. Revealing to the reader the Laws of Nature. Everyday physical, astral and spiritual phenomenon are explained, through the proper universal principle. The terminology is simple, yet profound, opening a door to the reality of the dynamics of everyday life.

The diligent reader, will be rewarded with the understanding of the causes of natural things. The understanding of Nature includes the understanding of one's Self and will reveal the causes of our errors. Thereby greatly alleviating our suffering and fears.

Path to Health
Author: Gerhard Hanswille
Date: late 1991-early 1992
ISBN 0-9693820-3-0
Format: probably hardcover

These writings contain the fundamental principles of health, and the use of herbs based on the laws of Nature and Classic Herbology. A very comprehensive work.